INTERNATIONAL SPACE STATION:
THE NEXT SPACE MARKETPLACE

SPACE STUDIES

VOLUME 4

Editor

Prof. MICHAEL RYCROFT

International Space University

Excellence in space education for a changing world

The International Space University (ISU) is dedicated to the development of outer space for peaceful purposes through international and interdisciplinary education and research. ISU works in association with a number of Affiliates (universities, research institutes, consortia ...) around the world and in partnership with space agencies and industry.

For young professionals and postgraduate students, ISU offers an annual two-month Summer Session in different countries and an eleven-month Master of Space Studies (MSS) program based at its Central Campus in Strasbourg, France. ISU also offers short courses and workshops to professionals working in space-related industry, government and academic organizations.

Independent of specific national and commercial interests, ISU is an ideal forum for discussion of issues relating to space and its applications. The network of alumni, faculty, guest lecturers, Affiliate representatives and professional contacts which characterizes the ISU Community makes it possible to bring together leading international specialists in an academic environment conducive to the exchange of views and to the creation of innovative ideas. ISU aims to promote productive dialogue between space-users and providers. In addition to the Annual Symposium, ISU supports smaller forum activities, such as workshops and roundtables, for constructive discussions which may help to chart the way forward to the rational international utilization of space.

INTERNATIONAL SPACE STATION: THE NEXT SPACE MARKETPLACE

Proceedings of
International Symposium
26–28 May 1999, Strasbourg, France

Edited by

G. HASKELL
International Space University

and

M. RYCROFT
International Space University

SPRINGER-SCIENCE+BUSINESS MEDIA, B.V.

A C.I.P. Catalogue record for this book is available from the Library of Congress.

ISBN 978-0-7923-6142-8   ISBN 978-94-011-4259-5 (eBook)
DOI 10.1007/978-94-011-4259-5

Printed on acid-free paper

ISU gratefully acknowledges the financial sponsorship provided by

The Boeing Company (Proceedings Sponsor)
DaimlerChrysler Aerospace AG, Space Infrastructure
European Space Agency
United Nations, Office for Outer Space Affairs

Table Of Contents

Acknowledgements

ISU acknowledges with thanks the advice and support given by the following people as members of the Program Committee:

B. Agrawal, Dept. of Aeronautics and Astronautics, Naval Postgraduate School, USA

O. Atkov, Faculty, International Space University

F. Becker, Dean & Vice-President for Academic Programs, International Space University

J.-P. Bombled, Space Utilisation Direction, Space Business Unit, Aerospatiale, France

J. M. Cassanto, President, Instrumentation Technology Associates, Inc., USA

P. Cohendet, BETA, Université Louis Pasteur, France

A. Eddy, Manager, ISS Commercialization, Canadian Space Agency, Canada

Y. Fujimori, Special Advisor, NASDA, Japan

S. Gazey, Vice-President Strategy & Business Development, DaimlerChrysler Aerospace AG, Germany

B. Harris, Vice-President Science & Health Services, and Chief Scientist, SPACEHAB, USA

R. Jakhu, Institute of Air and Space Law, Faculty of Law, McGill University, Canada

K. Knott, Head of Microgravity and Space Station Utilization Department, European Space Agency

S. V. Kulik, Senior Expert, Department of International Cooperation, Russian Space Agency, Russia

T. Kuroda, Corporate Chief Engineer, NEC Corporation, Japan

A. Nicogossian, Associate Administrator for Life and Microgravity Sciences and Applications, NASA, USA

H. Ripken, Coordinator for Space Station Utilization Preparation, German Aerospace Center, DLR, Germany

G. Rum, Italian Space Agency, Italy

N. Tolyarenko, Director, Master of Space Studies Program, International Space University

J. von der Lippe, Managing Director, INTOSPACE GmbH, Germany

J. Vaz, President, BRAZSAT, Brazil

G. Haskell, Vice President for Programme Development, ISU

Symposium Programme Committee Chair: **G. Haskell**, ISU

Symposium Convenor: **P. French**, ISU

Symposium Co-ordinators: **L. Chestnutt, E. Vossius**, ISU

Proceedings Editors: **G. Haskell** and **M. Rycroft**, ISU

Editorial Assistant: **L. Chestnutt**, ISU

Foreword

G. Haskell, Symposium Programme Committee Chair, Vice President, Administration and Programme Development, International Space University

e-mail: Haskell@isu.isunet.edu

M. Rycroft, Faculty Member, International Space University

e-mail: Rycroft@isu.isunet.edu

The theme of the fourth annual symposium arranged by the International Space University (ISU) was "International Space Station: The Next Space Marketplace". The Symposium covered this topic from the unique — interdisciplinary, international and intercultural — perspectives of ISU. It focussed on significant issues related to policy, innovative management, commerce, regulation, education and outreach rather than concentrating on engineering and scientific issues.

Although admirable progress has already been made in defining the utilisation of the International Space Station (ISS) in its early operational phases, what does the future hold? What important new applications will arise? What commercial opportunities may emerge? And how will the political, legal and financial hurdles be overcome, not to mention the technical challenges? The aim of the Symposium was to discuss such questions and draw out new ways of using the Space Station in the future.

Among the 120 attendees were members of the fourth Master of Space Studies class, young professionals and postgraduate students who are developing the Symposium's theme in their Team Project. Their comprehensive overview of the subject is presented as an Annex here. Their final report on the Team Project will be completed at the end of July 1999, and published separately.

To sum up, these proceedings of the Symposium are essential reading for all who may wish to use the International Space Station, and all those who plan to attend the ISS Forum 2000 in Berlin, Germany, in June 2000 — see http://www.estec.esa.int/ISSForum2000

Keynote Address

ISS: From Political to Scientific and Economic Preeminence

W. Kröll, H. W. Ripken, German Aerospace Center, 51147 Köln-Porz, Germany

e-mail: Walter.Kroell@dlr.de, Hartmut.Ripken@dlr.de

Abstract
The ESA Ministerial Conference of May 1999 made a great step forward in guaranteeing the European elements of ISS exploitation and utilization. Emphasis in this area has been put on public-private-partnerships and on industrializing the ISS exploitation. A strongly emerging industrial interest in ISS utilization is evident which not only reflects activities to promote the involvement of non-space industry in Germany, and by ESA, but also generally within the countries of the global ISS partnership.

The characteristics of industrialized space utilization in the future are reduced government involvement, and the acceptance of responsibility and risk by the private sector. The successful opening of a "space marketplace" depends on certain key elements — the identification and implementation of a well-defined set of utilization prerequisites, and the systematic development of new ISS business and research areas.

The ISU Symposium can help to foster the "Next Space Marketplace"; the results of the Symposium need to be forwarded effectively and must be introduced into the global space dialogue.

1. Introduction

The decision to build the International Space Station was primarily politically motivated. It is well known that scientists never requested a Space Station, and science alone does not provide a justification for the ISS. It is, however, equally clear that politics — as well as the space agencies involved in the construction of the ISS — firmly and determinedly require that the best possible use be made of the new space infrastructure. Within this frame, a perspective for the ISS must be drawn that encompasses multipurpose utilization, both by science and by industry. The course of action necessary to reach this goal is at present not very obvious; the ISU Symposium 1999 was conceived and implemented to help shed some light on this issue.

There is no automatic generation of a new marketplace. The goal needs to be actively pursued by all partners — by academia, industry and government. It must be an internationally coordinated effort that eventually will lead to a global marketplace without regional boundaries and without demotivating obstacles for the users.

1

G. Haskell and M. Rycroft (eds.), International Space Station, 1-6.
© 2000 *Kluwer Academic Publishers.*

2. A Vision of the ISS

Space infrastructure such as Spacelab or SpaceHab have offered the space community large payload resources (such as power and mass), combined with irregular and infrequent access for users. The Mir Space Station offered regular and frequent access, combined with very small payload resources. The International Space Station will offer researchers a fundamentally new capability — large resources and frequent access — enabling researchers to use the ISS as a powerful new "tool" for scientific purposes and, especially, for industrial utilization.

The ISS is to become an integral part of the infrastructure for research and technology development, by expanding, supplementing and enriching the classical ground-based facilities for research. Embedding ISS research in such a way ensures a higher acceptance in the science community and offers, through established mechanisms of peer review or merit evaluation, the prospect of maintaining a high standard of science or technology.

2.1 Utilization and Business Areas

As a goal, Germany and ESA aim at a "balanced utilization" of the ISS, allowing approximately equal shares for fundamental sciences, for applied research, and for industrial utilization. While high standards in science and an adequate number of proposals for scientific experiments already appear to be ensured in the early utilization phases, the quantity of good industrial proposals needs to be enhanced by special measures.

For such proposals, three main areas can be identified: (1) the "classical" area of the space industry (infrastructure, building of facilities and instruments); (2) an emerging area in the field of operations and logistics, and (3) the new area of utilization by the space industry and by non-space industry (mainly research and technology development for terrestrial and space applications).

2.2 Result of the ESA Ministerial Conference

The Ministerial Conference in Brussels (May 1999) provided important decisions on ISS and related issues, stabilizing the ISS utilization, exploitation and operation, while emphasizing the three business areas mentioned above. In general, the ESA Member States decided to develop a coherent European Space Strategy, and to build up a network of European "Centers of Competence".

Furthermore, an increased engagement of the private sector was called for, implying extended public-private-partnerships (PPP). PPP in this context is more than merely a co-financing: the partners commit themselves to common goals, concentrate each on their specific strengths, and mutually rely on each other. A PPP can only function if advantage and profitability are evident for each partner. An emerging public-private-partnership in the ISS area is visible in the operations and logistics of the Space Station.

Specific ISS-related results of the Ministerial Conference include continuing microgravity research and the new program element of a CRV (Crew Rescue Vehicle). As a measure directly impacting the ISS as a "space marketplace", the ESA Director General was requested to propose a concept for industrializing the ISS exploitation, to be delivered by March 2000.

3. ISS Utilization and Industrial Engagement

Keeping in mind recent developments in ISS partner countries, and especially the ESA Ministerial Conference, two general goals of ISS utilization can be identified: excellence in science, and utilization with commercial perspectives.

These are longterm goals; however, already in the current early utilization phase encouraging results are to be noted in Europe. In the scientific area, excellent proposals have been submitted, oversubscribing the available resources in Europe by a factor of more than 2. In particular, the researchers already include two Nobel laureates. In the industrial area, the emerging interest in ISS research is evident in numerous areas, ranging from metallurgy and robotics to cell biology and pharmaceutical drug design. The first commercial ventures with substantial investments by the private sector will be in signal transmission services and in high temperature superconducting telecommunications equipment. With these proposals already having been submitted today, a balanced participation of science and industry appears not to be impossible in the exploitation phase, after 2004.

3.1 German Efforts in Industrial Promotion

Increasing participation of non-space industry in ISS utilization implies "winning" paying customers and investors. This will not happen automatically, but requires active marketing of the ISS utilization opportunities. To this end, the

German Space Agency (German Aerospace Center, DLR), has initiated a project for the promotion of industrial users of the ISS.

Its prime objectives are (1) stimulating a wide-spread interest within the non-space industry by means of an "Information Service ISS", addressing not the general public but potential users and decision-makers, (2) establishing close contacts with National Associations of Industry and initiating concrete cooperations with these "mediators" to their member companies, and (3) establishing a system of "user coaching", providing a single point-of-contact for new users. Task (3) is emerging at present as a further example of PPP, bringing together, e.g., the industrial user service center, BEOS, and the German Aerospace Center, DLR. Existing User Centers in Germany, such as MUSC in Cologne and ZARM in Bremen, are integrated into these activities, as well.

3.2 ESA Efforts in Industrial Promotion

ESA is focussing its efforts on "Topical Teams". Currently there are 24 teams consisting generally of groups of people from academia and industry. Selected teams can continue their work in specific fields until flight-ready experiments are developed. While the German approach relies on wide-spread "marketing", ESA chooses to select a few, especially promising teams.

In the area of applied microgravity experimentation, the recent MAP (Microgravity Application Promotion) Announcements of Opportunity yielded an overwhelming response. Of 144 proposals, 61 were with the involvement of industry, with a total number of 163 companies being named.

In June 1999, the ESA User Information Center on ISS Utilization, located in ESTEC/Noordwijk, will be inaugurated, providing numerous information services to European experimenters including industrial users.

3.3 Towards Industrial Engagements

Both approaches, the selective ESA method and the broader German method, have already resulted in the first successful activities, where industrial companies are committing substantial funding. Clearly the status of ISS utilization is in a "Phase 1" stage, where experiment costs are paid partly by the private sector, while mission costs (transport, logistics, system operations) are paid by the public sector.

	Experiment Costs	Mission Costs
Phase 1	partly private	fully public
Phase 2	fully private	fully public
Phase 3	fully private	partly public

Table 1. Utilization Phases

The three phases of ISS utilization are indicated in Table 1; the goal is to reach Phase 3. Parallel with this development, the governments need to step back, need to reduce subsidies and hand over the initiative in the area of industrial utilization to the private sector. Companies wishing to participate in and to exploit the ISS utilization opportunities need to accept responsibility and risk, while aiming at making a profit and stabilizing — or even increasing — their market position.

A general principle emerges here: it is that there should be as much private activity as possible, and as little public activity as necessary. Examples of necessary government functions include legal, regulatory and supervisory functions.

3.4 The Task: Open the "Space Marketplace"

Current efforts and trends have been described. It is vital here and now to state the necessary boundary conditions for reaching the goal of a "Space Marketplace". The key elements can be formulated as follows:

- Frameworks for proprietary research by industry and for commercial services need to be established by the ISS partners

- Legal and regulatory aspects must be covered; they should be absolutely transparent to the users

- The potential ISS business areas need to be identified; they must be actively and systematically developed. Six areas can be named currently:
  - Technology testbed
  - Research, both public and private
  - Observational activities, e.g. Earth observations
  - Operations, logistics and commercial services
  - Education and outreach
  - Free market elements, such as advertising

- Prerequisites for ISS utilization by industry need to be identified and must be secured by the ISS partners:

- Regular and frequent access to ISS
- Transparent selection criteria; no standard (scientific!) peer reviews; industrial access also possible via national selection processes
- User-friendly access conditions and procedures
- Assured confidentiality and proprietary rights for industry
– Reliable schedules and costs.

All of these key elements are essential for risk and profit evaluation by industry; thus they constitute clear "go – no go" criteria for industrial involvement, and for the opening of a "Space Marketplace".

4. Conclusions and Outlook

It has been stated before, and it has become manifest in almost all elements of the industrial and commercial development of ISS utilization: "There is no automatic generation of a new space marketplace". It needs to be rigorously worked for by all the parties involved; programmatic and policy constraints need to be minimized. All "players" are asked to coordinate, cooperate and pool their resources — the private sector, national agencies, the European Space Agency, the international ISS partners, the European Union, and the United Nations.

The results arising from this Symposium must be collected, analyzed, processed and adequately disseminated. The Symposium will not be complete until the results and recommendations are introduced into the global ISS dialogue. One opportunity for this will be at the UN conference UNISPACE 3 in Vienna, in July 1999. Others include workshops of the space agencies, strategic and tactical discussions on a bilateral or multilateral basis, and the next Global ISS Utilization Conference to be held in Berlin, in June 2000.

In accepting their respective tasks, individual "players", the agencies, and industry can foster the necessary private sector engagement and thus help to make the International Space Station become the next space marketplace.

Session 1

Management of the Research and Development Process

Session Chair:

J. G. Vaz, BRAZSAT, Brazil

Brazilian Participation in the International Space Station, with an Emphasis on Microgravity Research

J.G. Vaz, Brazsat - Brazilian Commercial Space Services, Av. Dr. João Guilhermino, 261 suite 133, São José dos Campos, S.P. 12227, Brazil

e-mail: brazsat@aol.com

J.A. Guimarães, University of Rio Grande do Sul, Centro de Biotecnologia - UFRGS, Av. Bento Goncalves 9500 C.P. 15005, Porto Alegre 91501-970, Brazil

e-mail: Guimar@conex.com.br

Abstract

Brazil has various vigorous and well-established national programs, which are government supported, and which underpin future Brazilian utilization of the International Space Station (ISS). The particular opportunity for Brazilian participation in the ISS is for staff at government research centers and private companies to use its research and development facilities to develop a productive enterprise initiative including both the private and public sectors. Of particular importance is the opportunity offered for projects in the field of pharmaceutical products, an area that Brazil and the entire South American market is now developing and already one of the largest markets in the world.

Brazilian research agencies are excited about using the ISS and also upcoming Space Shuttle flight opportunities that are linked closely with ground research as a new way of doing research. Projects can be designed to develop new drugs, vaccines, protein pharmaceuticals, diagnosis kits and other products for the treatment, control and prevention of tropical and parasitic diseases, involving both government centers and the private sector.

1. Overview of the Brazilian Academic Network

The Brazilian Universities and the nation's science and technology activities are relatively recent. The University of São Paulo (USP), the largest and most scientifically oriented Brazilian University, was created only in 1934. Much of the country is still in an underdeveloped condition, due to the delay in establishing a strong academic sector. It was only in the last four decades that the need for changes in this situation became a matter of governmental concern.

A national program for graduate studies was officially organized at the end of the 1960s. This program was primarily designed and conceived as a route for increasing the qualified personnel capable of improving the quality of teaching and strengthening the research activities at universities and other national institutions. At the same time it was expected that the program would contribute to the technological development of the country by supplying it with well-trained scientists and technologists dedicated to research and

G. Haskell and M. Rycroft (eds.), International Space Station, 9-16.
© 2000 *Kluwer Academic Publishers.*

development, and meeting the needs of both public and private industrial sectors [Reference 1].

The Ministry of Education continuously supervises the program as a whole and its scientific output is evaluated biennially by special ad hoc peer review committees. During the last three decades this program has been consolidated and is functioning in several universities and research centers that are officially accredited to offer degrees at Master's and/or Ph.D. levels.

The scientific output of Brazil has increased continuously in both quality and quantity. Publication of scientific articles in the most acknowledged international journals and periodicals, as catalogued by the Science Citation Index, the Philadelphia-based Institute for Scientific Information [Reference 2], increased progressively in the period 1973-1998 as shown in Fig. 1.

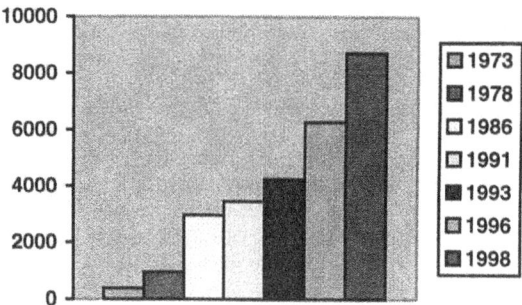

Figure 1. Number of Brazilian Papers Published [Reference 2]

Furthermore, in this period the growth rate of publications was twice the average rate. Highlights of these advances have been pointed out recently [References 3, 4, 6].

2. Key Areas of Research Successes in Brazil

As a consequence of this training program, Brazil has made an impressive impact in several technological areas [Reference 1]. Some key achievements include:

- technology for exploring petroleum resources in deep sea water
- cellulose and paper-mill industries, and woodland recovery
- design and construction of alcohol-propelled motor vehicles
- soil technology to develop large areas of once infertile land known as "cerrados" and "caatingas"
- advances in electronics and telecommunications systems
- computer science, especially software development and robotics
- sophisticated automation for the nationally-expanded network of Brazilian banks
- aircraft design and production
- launching of scientific satellites
- satellite-supported systems, including data collection platforms and remote sensing stations for monitoring forest fires, Atlantic rain forest climate, and agricultural and environmental forecasts
- advances in medicine, especially cardiac surgery
- scientific knowledge of insect-borne tropical diseases
- agribusiness, covering plant, animal and food technology, production and industrialization of soybeans, coffee, sugar cane, citrus juice and tropical fruits.

3. Characteristics of the Brazilian Industrial Sector

Despite these achievements, Brazil still lacks a strong technological and industrial sector capable of competing at the international level. There are several reasons for this. The national industrial sector neither invests in science and technology nor does it require research support. On the other hand, international companies located in the region usually prefer to transfer their own technology to their foreign branches. Nevertheless, some industrial sectors seem to be moving in the direction of a more cooperative effort to explore Brazilian opportunities from both economic and scientific viewpoints.

3.1 The Pharmaceutical Sector in Brazil

At present, a great opportunity exists for the pharmaceutical industry to be engaged in R&D in Brazil.   The sector benefits from several attractive advantages:

- Brazil is becoming a significant producer of pharmaceutical products; from 1992 to 1995, the annual rate of growth was approximately 30%
- Investments have reached a level of US $600 million
- The motivation for this boom is the size of the Brazilian market, the sixth largest market in the world, and capable of further growth
- The existence of the largest biodiversity environment in the world
- Several laws which stimulate the application of industrial funds for research
- A law which regulates the production and commercialization of biotechnological and other products
- A beneficial patent law.

4. Microgravity Research Opportunities for Drug Development

Projects for drug developments based on target proteins are now facilitated by microgravity facilities for growing pure protein crystals of uniform size in space.   Proteins are important, complex biochemicals that serve a variety of purposes in living organisms. Metabolic processes involving proteins play an essential role in our lives, from providing nourishment to fighting disease. In the past decade, rapid growth in protein pharmaceutical use has resulted in the successful application of proteins to insulin, interferon, human growth hormone, and tissue plasminogen activator.

Other potential applications include agricultural products and bioprocesses for use in manufacturing and waste management. Such research has attracted firms in the pharmaceutical, biotechnological, and chemical industries.

In response to these opportunities, Brazsat has teamed-up with the Center for Macromolecular Crystallography at the University of Alabama in Birmingham (CMC-UAB), USA.   This is one of the NASA Centers for the Commercial Development of Space (CDS) that form a bridge between NASA and private industry; it is developing methods for the crystallization of macromolecules in microgravity. The CMC-UAB has formed affiliations with companies that are investing substantial amounts of time, research, and money to develop protein samples for use in evaluating the benefits of microgravity. Protein structural information leads to the discovery and synthesis of complementary compounds that can become potent drugs specifically directed against the disease target. Structure-based drug design is a productive and cost-effective targeted drug development strategy.

The commercial applications developed using protein crystal growth have phenomenal potential, and the number of proteins that need to be studied is in the tens of thousands. Current research, with the aid of pharmaceutical companies, may lead to a whole new generation of drugs, which could help treat diseases such as cancer, rheumatoid arthritis, periodontic disease, influenza, septic shock, emphysema, aging, AIDS and tropical diseases.

5. Role of Brazsat in ISS

Being the only Brazilian Company working in the field of commercial space applications and utilization of microgravity, Brazsat was responsible for the major achievements that led to the Brazilian participation in the ISS; in 1998 Brazil became a partner in the ISS program. Brazsat is recognized by the United Nations Committee for the Peaceful Uses of Outer Space as an example of a space company working for the benefit of developing nations. In April 1997 Brazsat coordinated the first Brazilian Biotechnology experiment to fly in space, introducing to Brazil the concept of using the microgravity environment to enhance research.

One of the main justifications for Brazil's participation in the ISS program was the research benefit to be achieved from carrying out protein crystal growth experiments in the microgravity environment. Several diseases such as chagas, dengue and malaria affects millions of Brazilians and cost the government several billion dollars.

As a member of the International Space Station Program, Brazil will be contributing by supplying NASA with important hardware; NASA will allow Brazil to use the US segment of the ISS (0.25% utilization). Looking ahead to 2005, Brazsat is working to assist Brazilian scientists and industries to begin to use microgravity in order to gain the necessary expertise for when the Space Station utilization time is available. The key areas for research for the Brazilian scientific community are in protein crystal growth and agriculture research (plant growth). This type of research cannot be done using sounding rockets, since it requires a reliable access to space and long duration flights with very sophisticated hardware and scientific preparations.

Brazsat has already coordinated the first series of Brazilian experiments aboard the NASA Space Shuttle on the missions STS 83, 94, 84, 91 and 95. These flights in less than two years clearly demonstrate the continuation of services and also the excellent support to the Brazilian research community.

In October 1998, President Fernando Henrique Cardoso wrote an important letter to President Bill Clinton stating that the research being carried out by the Brazilian scientific community on the Space Shuttle was extremely important to Brazilians, being an important advance in the researches for both new drugs and agriculture research.

Brazsat has partnerships with other Brazilian aerospace industries and research agencies to develop hardware that would be used by Brazilian researchers aboard the ISS. Initial contracts have already been signed for the development, on a commercial basis, of several pieces of hardware tailored to support the scientific needs. As a result, they will generate fantastic opportunities to small Brazilian aerospace firms to design, build and test manned space flight hardware.

Brazsat's next initiative is to open, in October 2000, a Commercial Space Utilization and Education Center to serve the Brazilian and South American markets. Initial funding is already allocated and the construction of the center and implementation of the first phase should soon start. This center will include the direct involvement of several key players in the area of microgravity research, also Brazilian pharmaceutical and agriculture-related companies, and venture capital groups. The center will be called South American Microgravity Utilization and Integration Center – SAMUIC – and it will support the growing market in space research on a commercial basis without bureaucracy. The center will be strongly supported by existing partnerships with leading research centers in Brazil and in other countries, and it will coordinate any type of microgravity experiment ranging from sounding rockets to the ISS.

Some of the other important initiatives that Brazsat is actively participating in to move forward the space business in Brazil and to also promote the ISS program are:

- Building the necessary political support for microgravity services at different levels in the political system. Politics are important and related to space, and vice-versa. In Brazil funding for space research must be directly related to benefits gained by Brazilian society
- Building coalition support for space research to create funding for Brazilian research institutions and scientists to fund their microgravity research needs. Brazsat is working closely with several government agencies to create a well-defined funding structure for microgravity research, and also to support the ground segment of research both before and after the experiments are flown in space.

Brazsat has already organized two commercial space workshops. The third will be held in Rio de Janeiro in August 1999.

6. Other Key Players in Brazilian Microgravity Research

- AEB (Brazilian Space Agency) is the official space-related agency in Brazil, with the mandate to organize Brazilian space activities. It is newly created (1994) and has very limited budget. It created, in 1998, a microgravity program based on the utilization of Brazilian sounding rockets and the Brazilian allocation aboard the ISS

- BIO-AMAZONIA is a biotechnology center to be created by the Brazilian government as part of the Brazilian Government official Plan of Social Action. Brazsat has a Memorandum of Understanding to cooperate with this center in promoting the use of space as a resource for research

- EMBRAPA, the leading agriculture research agency in Brazil, will be flying proteins to target some of the most important agriculture related research that will have direct benefit to Brazilian agricultural production

- FAPESP (São Paulo State Research Foundation) funds researchers, universities and industries located in the State of São Paulo to perform microgravity research aboard several Space Shuttle missions

- FINEP, the funding agency for studies and projects closely tied to the Ministry of Science and Technology in Brazil, will be a key player in future microgravity research projects in Brazil

- FIOCRUZ, the leading research agency of the Ministry of Health and a center of excellence for research on tropical diseases and AIDS

- INPE (National Institute of Space Research) was responsible for the coordination of the first series of Brazilian microgravity experiments. INPE is a key player and its engineers are already fully trained by CMC-UAB to integrate manned space flight hardware aboard the Space Shuttle. The next step is jointly to develop science-related microgravity hardware in close cooperation with Brazilian industry

- LNLS (National Synchrotron Laboratory), located in Campinas (West of São Paulo), is the only synchrotron source in Latin America. Several Brazilian experiments that were aboard the Space Shuttle were analyzed here

- USP (University of São Paulo), participated already in several microgravity experiments in the area of protein crystal growth and obtained excellent results.

References
1. Guimarães, J. A and Humannn, M. C.: *Scientometrics 34*, pp.101-119, 1995
2. *Science Citation Index Institute for Scientific Information*, 1999
3. *Science Watch 267*: pp. 807-828, 1995

4.   Pulling together in Latin America, *Nature 398*: p. 353, 1999
5.   *Science 267*: 807-828, 1995
6.   Pulling together in Latin America, *Nature 398*: p. 353, 1999

BEOS – A New Approach to Promote and Organize Industrial ISS Utilization

B. Bratke, H. Buchholz, H. Luttmann, DaimlerChrysler Aerospace AG, BEOS, Postfach 286156, 28361 Bremen, Germany

e-mail: burkhard.bratke@ri.dasa.de,   henning.buchholz@ri.dasa.de, helmut.luttmann@ri.dasa.de

H.-J. Dittus, Universität Bremen, ZARM, Am Fallturm, 28359 Bremen, Germany

e-mail: dittus@zarm.uni-bremen.de

D. Hüser, OHB-System GmbH, Universitätsallee 27-29, 28359 Bremen, Germany

e-mail: hueser@ohb-system.de

Abstract

In order to develop and to market innovative services and products for the operation of the ISS and its utilization, three players have teamed up together and established an entity called BEOS (Bremen Engineering Operations Science). The team is made up of DaimlerChrysler Aerospace, OHB-System and ZARM, the Center of Applied Space Technology and Microgravity at the University of Bremen.

It is the aim of BEOS to represent a competent industrial interface to potential ISS users from the space and non-space industries. In this effort BEOS is supporting and supplementing the activities of the space agencies, especially in the field of industrial and/or commercial ISS utilization. With this approach BEOS is creating new business opportunities not only for its team members but also for its customers from industry. Besides the fostering of industrial research in space, non-technical fields of space utilization like entertainment, advertisement, education and space travel represent further key sectors for the marketing efforts of BEOS.

1. Introduction

During the last decade several sectors of the space industry have changed from pure receivers of public support into high growth business fields. Especially in the satellite telecommunications segment this transformation is visible, and navigation and Earth observations are following closely behind. In this context the utilization and operation of the International Space Station (ISS) are discussed with respect to their business potential. In general there are three major areas for the creation of business revenues:

*   Carrying out ISS operations
*   Industrial research
*   Entertainment, advertising and education.

G. Haskell and M. Rycroft (eds.), International Space Station, 17-22.

In order to develop and to market innovative services and products for the operation of the ISS and its utilization, three organizations, whose home locations are Bremen, Germany, have set up an entity called BEOS (Bremen Engineering Operations Science). The three players comprise DaimlerChrysler Aerospace, OHB System and ZARM, the Center for Applied Space Technology and Microgravity at the University of Bremen (see Fig. 1). The combination of a major international corporation, a medium sized enterprise and a research institute is expected to provide an excellent base for the development and service of the ISS operations and utilization market.

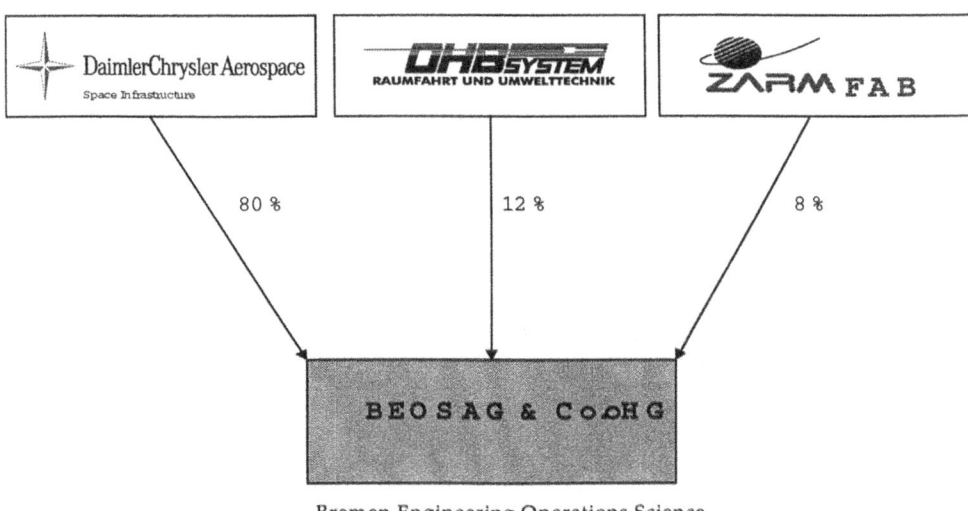

Bremen Engineering Operations Science

Figure 1. The relative involvement of each of the three partners in BEOS

2. The BEOS Partners

In this section the founding members of BEOS and their experiences in the development, operation and utilization of microgravity research facilities are briefly summarized.

2.1 DaimlerChrysler Aerospace

The first BEOS partner, with a share of 80% and the main shareholder, is the space infrastructure department of DaimlerChrysler Aerospace (DASA).

DASA Space Infrastructure in 1997 had annual sales of the order of US $ 600 million and a work force in Bremen of around 1200 employees. Under its former name ERNO Raumfahrttechnik this department developed and operated the TEXUS sounding rocket program for microgravity research in the 1970's. As prime contractor the development of Spacelab was the next milestone; its first flight took place in November 1983. In the following years until the retirement of Spacelab in 1998, ERNO/DASA built facilities and research equipment for several missions, e.g., the German missions D1 and D2. In addition support was given to the space agencies and researchers before, during and after the missions with respect to facility operations, astronaut training and experiment logistics. Further examples of the involvement of ERNO/DASA in the space infrastructure segment for microgravity research was through the development and operation of the space pallet satellite (SPAS, flights in 1983, 1993 and 1994) and the free flyer platform EURECA (flight in 1992/93). In 1996 DASA was selected as the prime contractor for the development and integration of the European ISS module Columbus. Additional participation of DASA in the ISS program is in the Automated Transfer Vehicle (ATV), the European Robotic Arm (ERA), the Crew Return Vehicle (CRV), the Data Management System (DMS), the Integrated Cargo Carrier (ICC) and a maintenance unit called Inspector.

2.2 OHB-System GmbH

The second BEOS partner is OHB-System GmbH, the core company of the Fuchs Gruppe, a medium size company which has a share of 12% in BEOS. In 1997 OHB-System had an annual turnover of about US $ 20 million and 125 employees. Its activities in the space sector are widespread: they range from small satellites, re-entry technology, sensors for environment observation and mobile satellite communications terminals to microgravity systems and experiment facilities, space subsystems and ground support equipment.

In the field of microgravity research payloads, OHB-System participated in 6 Space Shuttle missions and 2 Mir missions. OHB-System is contributing to the ISS program by designing, manufacturing and testing the Pre-Integrated Columbus APM (PICA) system harness and Mechanical Test Support Equipment (MTSE). For Node 2 and 3 of the ISS it is responsible for the secondary structure and the harness complex. It has already delivered 16 International Standard Payload Rack (ISPR) ground models to ESA/ESTEC. OHB-System is also involved in the development of the Automated Transfer Vehicle (ATV). In the ISS payload segment OHB participates in the development of the BIOLAB, the Modular Cultivation System (MCS), the Fluid Science Lab (FSL); it is prime contractor for the European Drawer Rack (EDR) and the European Physiology Module (EPM).

2.3 ZARM Drop Tower Operation Business Unit

ZARM is the third partner of BEOS, with a share of 8%. ZARM is a research institute of the University of Bremen with about 80 researchers, scientists and engineers working there. The main areas of research are space technology, aerodynamics, gravitational physics, combustion, hydrodynamic stability, rotational fluids, interface phenomena and ferrofluids. ZARM is world renowned for being the operator of a drop tower with a height of 140 m, which provides 4.74 s in microgravity conditions.

The role of ZARM with respect to the ISS program is in scientific research in the fields of combustion, fluid science and ferrofluids. In these areas ZARM is participating in ESA topical teams, which are the core elements of the ESA Microgravity application program. These teams have been set up in order to identify and develop links between basic research and industrial applications in the different fields of microgravity research.

2.4 Partner Summary

Summarizing the BEOS partners' experiences and know-how in the field of space infrastructure operations and utilization, the following commonalities and supplementary assets can be identified. What the three partners have in common are experience in manned space flight, participation in the ISS program and the same location in the city of Bremen. They complement each other in their size and corresponding assets. DASA as a major international corporation gives power and stability, OHB-System as a medium sized enterprise offers flexibility and innovation, while ZARM stands for very well recognized scientific research and expertise. Looking at this combination it is clear that BEOS has the right starting conditions for performing a successful job.

3. BEOS Objectives

It is the aim of BEOS to make use of synergistic effects of the partners' competences in the operation and utilization of space infrastructure. Based on this a major effort will be spent on an innovative opening up of new markets for manned space flight. Here BEOS wants to represent a competent industrial interface to potential ISS users from the space and non-space industries. In parallel, the institutional side will be addressed in order to shape future space programs through discussions between science and industry. In this field BEOS is supporting and supplementing the activities of the space agencies, especially in the field of industrial and/or commercial ISS utilization.

A BEOS objective with a geographic background is to promote Bremen as a center for manned space flight and microgravity research. This goal will be achieved not only for the business sector but also it is hoped that the ISS activities will be integrated into the PR and tourist program of the city of Bremen.

Becoming more specific, a set of tasks for BEOS will be to:

- Trigger a general willingness within non-space industry segments to think seriously about ISS utilization
- Develop products and services to increase the attraction for industrial ISS operations and utilization
- Develop procedures and sequences to optimize ISS operations and utilization, especially for users from non-space industries
- Prepare adequate administrative boundary conditions for industrial ISS users in cooperation with the space agencies
- Attract industries to use already existing space infrastructure (drop tower, parabolic flights, MIKROBA, TEXUS system, Spacehab).

Besides the fostering of industrial research in space, non-technical fields of space utilization such as entertainment, advertisement, education and space travel represent further key sectors for the BEOS marketing efforts. As an example for this sector, cooperation with Charles Wilp in the conception of the art module MICHELANGELO is mentioned.

4. BEOS Products and Services

The anticipated products and services of BEOS are derived from the partners' business segments and the aforementioned objectives and goals. The products in general consist of:

- Facilities, systems and/or instrumentation for ISS operations, utilization and maintenance (space segment)
- Communications and data link between user and project
- Test facilities
- Ground facilities
- Software for data processing, display and evaluation
- Simulation and virtual reality tools.

As for the services which BEOS will offer, these include:

- Consultancy and support preparing, carrying out and financing projects
- Handling project logistics

- Providing communications services
- Supporting crew training
- Support operations and maintenance for the Columbus Orbital Facility (COF).

It is our understanding that these products and services will enable BEOS to provide an end-to-end service for any customer who is planning a project on or with the International Space Station.

5. Outlook

With this approach, BEOS is creating new business opportunities not only for its team members but also for its customers from industry. Therefore it is the vision of the BEOS team that the operation and utilization of the ISS have a similar business potential as the space industry sectors mentioned at the beginning. This vision includes the ISS or its successor being operated and utilized completely by the private sector. Essential for this vision is a more efficient and cheaper space transportation system that provides access to space on a regular "train-like" basis, e.g. one flight per week for cargo as well as for astronauts/researchers/tourists. At the same time the space agencies will need all their manpower to make sure that the first manned mission to Mars is a success.

Protein Crystallography Services on the International Space Station

M. Harrington, T. Bray, W. Crysel, L. DeLucas, J. Lewis, University of Alabama at Birmingham/Center for Macromolecular Crystallography, MCLM 262, Birmingham, AL 35294-0005, USA,

e-mail: harringt@cmc.uab.edu,   bray@cmc.uab.edu,   crysel@cmc.uab.edu, delucas@cmc.uab.edu,   lewis@cmc.uab.edu

T. Gester, T. Taylor, Diversified Scientific, Inc., 2800 Milan Court, Suite 381, Birmingham, AL 35211, USA

e-mail: tgester@dsitech.com,   taylor@cmc.uab.edu

Abstract
The Center for Macromolecular Crystallography (CMC) has performed protein crystal growth experiments on more than 30 U.S. Space Shuttle missions. Results from these experiments have clearly demonstrated that the microgravity environment is beneficial in that a number of proteins crystallized were larger or of higher quality than their Earth-grown counterparts. These microgravity results plus data from a variety of other investigators have stimulated various space agencies to support fundamental studies on macromolecular crystal growth processes. The CMC has devoted substantial effort toward the development of dynamically controlled crystal growth systems, which allow scientists to optimize crystallization parameters on Earth or in space. This capability plus the CMC's experience in protein structure determination and structure-guided drug development have attracted partnerships with a number of pharmaceutical and biotechnology companies. The CMC is currently designing a complete crystallographic laboratory for the International Space Station. This facility will support a variety of crystallization hardware systems, an X-ray diffraction rack for crystal characterization or complete X-ray data set collection, and a robotically controlled crystal mounting system with cryo-preservation capabilities. The X-ray diffraction rack and crystal harvesting/cryo-preservation systems can be operated with minimal crew time via telerobotic and/or robotic procedures. The CMC, in conjunction with its spin-off company, Diversified Scientific, Inc. (DSI), is currently marketing crystallography services, which include crystal growth, structure determinations and structure-guided drug development.

1. Introduction

X-ray crystallography is a tool used by scientists to determine the three-dimensional structure of proteins or other macromolecules. From a basic research perspective, the structural information obtained enables researchers to see the mechanisms by which enzymes, receptors, hormones, and other macromolecules function in biological systems. From a commercial standpoint, this information has become invaluable in the development of new pharmaceuticals. Using a technique known as structure-based drug design, the protein structural information allows for new pharmaceutical agents to be designed that interact with specific sites on the protein molecule of interest. This

G. Haskell and M. Rycroft (eds.), International Space Station, 23-30.
© 2000 *Kluwer Academic Publishers.*

technique is being applied to the search for new drugs to combat both chronic and infectious diseases. The growth of large, well ordered protein crystals is a critical part of the application of X-ray crystallography for molecular structure determinations. The quality of the crystal has a direct impact on the quality of the diffraction pattern produced by an X-ray diffraction system and ultimately on the atomic resolution that can be determined from these diffraction images. Thus, the quality of the protein crystal is critical to the successful determination of a protein structure.

2. Protein Crystal Growth in Microgravity: Research and Benefits

2.1 Early History

Protein crystal growth is a relatively new science that often relies on understanding unique relationships of crystal growth parameters for each particular protein. There are many variables that affect the quality and repeatability of experimental results. In 1985 the Center for Macromolecular Crystallography (CMC) proposed to NASA flying protein crystal growth experiments on board the Space Shuttle with the goal of gaining a better understanding of the molecular events involved in macromolecular crystal growth. The hypothesis proposed was that removing the influence of gravity would affect crystal growth in three ways. First, the minimization of buoyancy induced convective flows would lead to a slower, more consistent crystal growth and diminish the inclusion of impurities. Secondly, the use of semi-containerless crystal growth would minimize potential nucleation sites, thereby leading to fewer, but larger, crystals. Finally, crystal sedimentation, which invariably occurs as protein crystals form, would be eliminated [Reference 1]. In that same year, the first of four flights occurred using simple proof-of-concept hardware developed in conjunction with Marshall Space Flight Center (MSFC).

2.2 Commercial Space Centers

The Commercial Space Center program was created by NASA to establish a consortium of universities and industrial partners to conduct focused research that would foster space commerce. NASA's concept was to do so by advancing broad domains of research endeavors, removing many of the practical impediments previously associated with space access, and developing a technology database. Based on promising results from the first proof-of-concept flights, the CMC was approved by NASA and established in 1985 as a Commercial Space Center. The focus of the CMC is to understand the structure and function of macromolecules, especially as they apply to biological processes and drug design. With support from industry, NASA, and other governmental agencies, the CMC has developed a multifaceted program that covers an entire spectrum from protein isolation and purification, crystal growth, structure determination, lead compound design and drug development (medicinal and structure-guided combinatorial chemistry). As part of this program, the Center relies on both ground- and space-based crystal growth experiments to achieve its mission goals. The benefits realized from the Commercial Space Center funding which NASA provides can be seen in the form of advances in research and development, as well as improvements in industrial competitiveness through education, commercialization, and economic return. The CMC to date has solved over 35 new protein structures, published over 400 papers, and received/applied for 19 patents. In 1998 the Center had active collaborations with 15 industrial partners, 37 academic institutions, and 5 government agencies. The CMC has continued to attract non-NASA funding from a number of other sources including industry foundations and other governmental agencies. For FY 98, the CMC's cash leverage ratio for the NASA Center Grant Funding was 1:17. For cash plus "in-kind" support, the ratio was 1:26. In addition, the CMC has created three successful spin-off companies; the most mature is Biocryst Pharmaceuticals, Inc., a publicly traded pharmaceutical company with a valuation of US $ 115 million (as of 26 April 1999).

2.3 Summary of Results from Microgravity

With the STS-95 mission, the CMC has flown 52 experiment systems on 34 Space Shuttle flights. Twenty-two of the experiment systems were flown under NASA's science code (Code UG), 29 were flown under NASA's commercial code (Code UX), and one was flown by SPACEHAB, Inc. (SHI) by leasing the experiment hardware from the CMC. A large co-investigator group consisting of scientists from the academic, government and commercial communities utilized these NASA/CMC flight opportunities. The CMC acts as an interface for each co-investigator to provide the scientist with a simplified and less intimidating process for accessing space. A large body of data has been collected from these experiments and the beneficial effects of microgravity have been summarized in a recent paper [Reference 2]. In addition, other NASA-sponsored flight investigators have conducted extensive microgravity crystal growth research, using his/her own unique hardware systems to obtain successful results. Two independent research groups have recently confirmed that protein crystals grown in microgravity are more perfectly arranged and produce better X-ray diffraction data than similar crystals grown on the ground [References 3 and 4]. The commercial implications of protein crystal growth in microgravity are not hypothetical, as crystals grown in space have played a role in new drug development for several diseases [Reference 2]. Examples include a new treatment for complications resulting from open-heart surgery, a broad-spectrum antibiotic, and new long-acting formulations of insulin. There are two primary concerns that exist when performing research or commercial activities relating to protein crystal growth on board the Space Shuttle. The first relates to the crystal growth time available. Protein crystals appear to grow more slowly in a microgravity environment [Reference 4]. For a large number of proteins, typical Space Shuttle missions are not of sufficient duration to allow sufficient time for crystals to reach a usable size. The other concern relates to the frequency of Shuttle flight opportunities. Typically, in a ground-based laboratory, an investigator must perform iterative experiments to optimize the crystallization conditions. The limited flight opportunities hinder this iterative process when performing microgravity research. Limitations such as these have prevented the commercialization of protein crystal growth services in microgravity. Investigators have long recognized these deficiencies; this is why they support the development of the International Space Station (ISS).

3. Protein Crystal Growth Microgravity Hardware Systems

3.1 Hardware Overview

The evolution of protein crystal growth systems has paralleled the increase in our understanding of fundamental crystal growth processes. The CMC's two primary areas of emphasis in each successive design involved increasing the density of experiment chambers and/or improving the probability of success in a given chamber via control of important crystallization parameters. While these improvements are important for the advancement of the science, they are an imperative for protein crystal growth to be commercially viable on ISS. The CMC realized this critical need early and in 1990 began to develop an in-house engineering capability to support the design, analysis, fabrication, assembly, testing, and mission operations for the experiment systems. Since that time the staff has grown to 48 engineers and technicians drawn from a variety of NASA contractors, including Boeing, Fairchild, McDonnell Douglas, Rockwell, Teledyne Brown, TRW, and Wyle Laboratories. The 1650 m^2 facility operated by the Engineering Group includes 280 m^2 of laboratory space and a clean room. It operates under a NASA certified quality program. Since 1990, the Group has developed over 11 experiment systems and has supported 30 of the 34 CMC protein crystal growth Space Shuttle flights.

3.2 Existing Experiment Systems

Following the four proof-of-concept flights, MSFC contracted the development of a system more suited for the performance of formal science experiments. The system designated the Vapor Diffusion Apparatus (VDA) flew for the first time on STS-26. Since then, the CMC has developed a number of experiment systems under grants or contracts for both NASA Codes UX and UG, respectively. Each system has capitalized on experiences from its predecessor in the areas of performance and experiment density. The number of individual experiment chambers has grown from 60 in the original VDA system to 128 in the present Commercial Vapor Diffusion Apparatus system. The Protein Crystallization Facility, first flown on STS-37, is a temperature-gradient system that has also undergone several upgrades on successive missions. The Protein Crystallization Facility with Light Scattering was the first CMC-developed system that included experiment diagnostics which allowed for real-time adjustments of experiment conditions by a crewman.

3.3 Dynamically Controlled Protein Crystal Growth System

The Dynamically Controlled Protein Crystal Growth (DCPCG) system is a new device under development at the CMC for use on the ISS as well as in research and commercial labs on the ground. The system can be configured to dynamically control either solution concentration or temperature. Using information from non-invasive diagnostics, active control of these parameters can, in real time, affect the supersaturation condition of the protein solution (for both pre- and post-nucleation phases). This novel technology is a departure from crystal growth methods normally used in laboratories. Traditionally, investigators force a protein solution into supersaturation by use of a precipitating agent. The vapor diffusion rate and final protein concentration is predetermined, thereby preventing intermediate control based on diagnostic information. Experiments using temperature control were similar in that a predetermined temperature profile was run for a given experiment. Iterative experiments are required for both of the conventional systems. Using diagnostic data, the DCPCG system adjusts the experimental conditions in real time to optimize crystal growth, thereby reducing the quantity of often-precious protein required. More importantly, the laser diagnostic capability provides pre- and post-nucleation control, which can significantly improve the experimental results. The initial reaction from crystallographers in both commercial and research labs has been extremely positive and the technology has been licensed to Diversified Scientific, Inc. (DSI), a CMC spin-off company.

3.4 High Density Protein Crystal Growth System

The High Density Protein Crystal Growth (HDPCG) system is being developed at the CMC for commercialization activities planned for the ISS. The system has 1008 experiment chambers designed to be removable from the growth system and placed in an appropriate facility for on-orbit sample removal and analysis. SHI has recently leased this system along with the technical support team for an upcoming SPACEHAB mission planned for STS-107. It is felt that the experiment density is approaching a point where the costs associated with flying a single locker equivalent can be offset with service charges that the market is willing to bear. SHI has established a team of international marketing representatives to market this system.

4. X-ray Analysis of Protein Crystals on ISS

4.1 Rationale

Macromolecular crystallographers are anticipating the opportunity to perform experiments on ISS because of the benefits provided by more flight opportunities and longer durations for crystal growth. However, these new opportunities are also creating several concerns. The time duration between flights to ISS is one issue. Most protein crystals are labile and will eventually begin to degrade. Also, commercial users will not rely on ISS as a resource if long time delays are experienced before crystals are returned. Another concern relates to the fragility of many protein crystals. The CMC has occasionally observed crystals that appear to have been damaged by loads associated with the return of the Shuttle to Earth and/or with post mission transport and handling. It is an unavoidable risk associated with protein crystal growth on the Shuttle. A method of performing crystal cryopreservation and/or X-ray analysis (for crystal quality determination and complete data set collection) on board ISS would eliminate these concerns.

4.2 X-ray Crystallography Facility

In 1996, the CMC completed a study for NASA Code UX regarding the feasibility of developing a facility for performing X-ray crystallography on-board ISS [Reference 5]. The study found the concept feasible with the exception of three technologies that would require advancement. The first was the X-ray source, which would require a reduction in mass, volume, and power to operate on ISS. A typical system in a laboratory weighs over 1800 kg and uses up to 10,000 W of electric power. The second was the need to select, harvest, mount, and snap freeze protein crystals robotically. Due to the structural complexity and fragility of the crystals, the only viable method relies on the dexterity of the human hand. The third was the X-ray detector system, which must not rely on hazardous coolants. By focusing resources on certain emerging technologies, the study indicated that the state of the art could be advanced such that the development of a facility for the ISS could be considered feasible. NASA funded the CMC to progress the identified technologies and develop a proof-of-concept laboratory system. The CMC, along with the firms in possession of the technologies identified in the feasibility study, have developed and demonstrated an operational laboratory version of the system, designated the X-ray Crystallography Facility (XCF). The planned facility can be integrated into one International Standard Payload Rack (ISPR) and be operated either by the crew or tele-robotically from the ground. The crystal mounting system is designed to harvest crystals robotically from the HDPCG (or equivalent) chamber blocks and mount crystals (selected by scientists on the ground) for cryo-preservation or X-ray analysis. The power and weight requirements of the X-ray source were reduced to 30 W and 23 kg.

5. Commercialization of Protein Crystallographic Services on ISS

5.1 Market surveys

There have been three market surveys performed to assess the demand for XCF services. The first was commissioned by the CMC and SHI to look at the potential market for the XCF among pharmaceutical companies and other potential users. Although specific information is still considered proprietary, general results indicated that an initial shakedown period of the system and creative pricing concepts would be necessary to "incentivize" early use of the system until it develops a performance track record. After proof of operation, firms would be willing to pay for protein structures not resolvable on the ground. The price that a given customer would be willing to pay varied widely depending on the size of the firm and the relative criticality of the protein in question. The second, a survey of NASA's protein crystal growth Principal Investigators, was performed by NASA's science code, Code UG. A majority of respondents ranked the need for the ability to perform X-ray diffraction on ISS as "desired" or "highly desired". Also, a majority of respondents ranked the need for cryopreservation of macromolecular crystals on ISS as "highly desired" or "mandatory". The final survey was performed by the CMC and focused on the international science user community. Of the 27 respondents surveyed (from academic and industrial sectors), 10 stated "definitely" and 10 "probably" that they would use the capabilities of the XCF. When asked if they would pay to use the capabilities of the XCF, 16 said "yes" and 11 said "no".

5.2 First Committed Customer

Diversified Scientific, Inc. is basing its future on the ability to utilize both ground and space laboratories to solve protein structures for its customers. It is this combination of ground and space services that is attractive to industry, as it maximizes the chance of providing the customer with high quality crystals. For the ground activities, DSI has licensed CMC technology to build and operate Dynamically Controlled Protein Crystal Growth systems. For proteins where these systems do not yield the data needed by the customer, DSI will use microgravity, the HDPCG system, and the XCF. Based on this plan, DSI considers itself an XCF customer and will use a combination of ground and microgravity services to achieve its business goals.

6. Conclusion

Market data and the experience of the CMC clearly indicate that a market for commercial protein crystal growth and utilization of the XCF exists. Based on this information, NASA requested the CMC to perform a business analysis of the XCF as a privately funded business. Cash flow analysis indicated that revenue streams would be large enough to cover operating costs, but would not allow recovery of the capital investment to build the XCF. Thus, a potential investor would have to desire a return other than revenue on its capital investment. Pharmaceutical companies have been approached, as they would be considered a primary benefactor of the facility. Although they have shown interest in paying a fee to use the XCF, it was evident that they are in the business of developing new pharmaceuticals, not machines and facilities. The remaining investor with the potential to meet the previously defined criteria is a government agency. NASA is working with the CMC to explore commercial and government options for funding the flight article. The XCF is manifested on UF-5, currently scheduled for launch in 2003.

Acknowledgments
The authors would like to acknowledge the crystallographers who have used microgravity to leverage their research for the betterment of humankind; the engineers and scientists at Bede Scientific Instruments Ltd., Bruker AXS, Inc., Oceaneering Space Systems of Oceaneering International, Inc., and the UAB/CMC as the technical advancements made in their respective fields have allowed for the removal of all technical barriers to placing an X-ray crystallography facility aboard the ISS; and to NASA for its vision and funding in support of these endeavors.

References
1. DeLucas, L. J., Smith, C. D., Smith, H. W., Senadhi, V-K., Senadhi, S. E., Ealick, S. E., Carter, D. C., Snyder, R. S., Weber, P. C., Salemme, F. R., Ohlendorf, D. H., Einspahr, H. M., Clancy, L. L., Navia, M. A., McKeever, B. M., Nagabhusan T. L., Nelson, G., McPherson, A., Koszelak, S., Taylor, G., Stammers, D., Powell, K., Darby, G. and Bugg, C. E.: Protein Crystal Growth in Mirogravity, *Science, Vol. 246*, pp. 651-654, 1989
2. Moore, K. M., Long, M. M. and DeLucas, L. J.: Protein Crystal Growth in Microgravity: Status and Commercial Implications, *CP458, Space Technology and Applications International Forum–1999*, edited by M. S. El-Genk, pp. 217-224. The American Institute of Physics, New York, 1999
3. Ferrer, J-L., Hirschler, J., Roth, M. and Fontecilla-Camps, J. C.: *ESRF Newsletter*, pp. 27-29, 1996
4. Snell, E. H., Weissgerber, S., Helliwell, J. R., Weckert, E., Hoelzer, K. and Schroer, K.: *Acta Crystallographica*, pp. 1099-1102, 1995
5. The University of Alabama at Birmingham-Center for Macromolecular Crystallography: *X-ray Crystallography Facility Phase A Executive Summary Report*, Document Number PCG-D-0030, March 31, 1996

Commercial Protein Crystallisation Facility: Experiences from the STS 95 Mission

W. Lork, G. Smolik, J. Stapelmann, DaimlerChrysler Aerospace (DASA), Dornier GmbH Space Infrastructure, 88662 Friedrichshafen, Germany

e-mail: wolfram.lork@ri.dasa.de,   georg.smolik@ri.dasa.de, juergen.stapelmann@ri.dasa.de

Abstract

In order to contribute to paving the way for the commercial utilisation of space in the area of protein crystallisation, DASA took the initiative, invested its own money and built a small crystallisation facility (CPCF) with support from the German Aerospace Center DLR, which financed the first mission. CPCF was integrated in an incubator, provided by the US company ITA.

The main features of the CPCF programme are presented; special emphasis is placed on procedural aspects, with particular respect being paid to the needs of industrial users. Special topics, i.e. confidentiality, the acceptance of procedures for toxic materials on the one hand and the typical basic requirements of commercial users on the other hand are addressed on the basis of this experience.

1. Introduction

Protein crystallisation and the subsequent X-ray diffraction analysis of the crystals are the standard analysis method in today's biochemical laboratories in universities and in industry. But more than 50% of the attempts to create crystals of sufficient size and quality have failed, and it is then not possible to determine the spatial structure of the protein.

Crystallisation of proteins in space has delivered crystals of higher quality and of larger size than in ground-based laboratories in many cases, as the crystallisation process is not disturbed by gravity-driven effects in the protein solution. The details of the crystallisation process and its laws are still topics of fundamental basic research. The investigation of crystal growth processes is one objective for such research in space, the other being the hope of achieving larger and better crystals for structural analysis, if this cannot be achieved on the ground.

As this kind of research is of importance not only for academic research but also for the pharmaceutical industry, protein crystallisation is one example of the commercial use of space. While, up to now, most of the research done in space was funded by public money and was judged by the excellence of the science, it is planned in the future to consider more application-oriented research, with the perspective of partial private funding of the space programmes. However, for these types of experiments it is no longer the excellence of the science, but the commercial interest of the company, which has

31

G. Haskell and M. Rycroft (eds.), International Space Station, 31-39.

to determine the character of these research projects. As this has not been the case very often in the past, it is necessary to perform pilot, or pathfinder, projects in order to determine the new rules. Such a pilot project was performed in November 1998 by DaimlerChrysler Aerospace (DASA) with support from the German Aerospace Center DLR. Pharmaceutical companies performed protein crystallisation experiments in a new environment, which was especially tailored to the needs of commercial companies. Some major experiences are outlined in this paper.

2. Protein Crystal Growth: Commercial World versus Academia World

2.1 *Protein Crystallisation*

Proteins are essential for life. The main functions of proteins in living organisms are structural support tasks and catalytic functions (enzymes). Especially for the functional understanding of the enzymatic functions, it is necessary to know the chemical and spatial structure of the protein molecules, which are different for each of the many thousands of enzymes controlling the biochemical functions in the human body.

An understanding of the enzyme function is a helpful prerequisite to developing new drugs. While in the past most drugs were identified by trial-and-error methods, including big screening programmes, the trend is now moving towards "smart" drugs, tailored to a specific purpose and with minimum side effects. This kind of "drug" engineering requires even more knowledge of the exact protein structure in order to design a drug properly fitting into the respective sites of the target protein.

The primary structure of proteins is determined by the sequence of amino acids, which are the building blocks of proteins. This sequence can be identified by chemical analysis methods, which are fully automated. The secondary structure of proteins (e.g. a helix) is given more or less by the primary structure, as it is known (from the amino acids' properties), how the chain will order itself. The tertiary structure describes how the chains form a three-dimensional structure. Although the basic laws of these structure formations are known, it is not possible to calculate them to a sufficient level of detail. The structure is determined experimentally by investigating the X-ray diffraction patterns of protein crystals. It is interesting to know that it is possible to derive the *in-vivo* structure of a protein in the living cell from the structural analysis of this protein in its crystalline form.

Although protein crystallisation and the subsequent X-ray or synchrotron beam diffraction pattern analysis is routine work, more than half of the attempts fail. The reason for this is that, for many proteins, it was not possible to crystallise them at all or the crystallisation ended in crystals which were too small or of insufficient quality for analysis. In those cases, the protein structures are still unknown.

2.2 Protein Crystallisation in Space

About 15 years ago, the first attempts at protein crystallisation in space were made; it was shown that crystals grown under microgravity conditions are larger and of better quality than those grown under similar conditions on the ground. Since these early days of experimentation under μg-conditions, numerous experiments have been performed in space, worldwide. All typical growth methods have been tested, many growth facilities developed and these are still being improved. Today we can state that the benefit of crystal growth in space is scientifically accepted; however, there are still a number of open scientific questions which need further investigations of the growth processes.

Industry is also carrying out protein crystallisation in space, often in teams with academic researchers for more general research topics. The other topic is in the competitive area, when the goal is to crystallise a particular protein of commercial relevance. In those cases industry is very reluctant to disclose information, but especially in those cases the benefits of using space may be very high. Therefore it is necessary to differentiate between the two application fields.

2.3 Academic and Pre-competitive Commercial Research

Both academic and pre-competitive research can be discussed together, as they are quite similar. Emphasis in this field is on elucidating the crystal growth process of proteins. For academic researchers it is important to understand the growth processes to extend knowledge. For the commercial world this research is of importance, as so many crystallisation attempts are failing, which has a significant impact on success.

This academic and pre-competitive type of research is therefore characterised by:

- Analysis of crystal growth mechanisms
- Optimisation of crystal growth (on the ground)
- Use of diagnostics in the experiments (observation, interferometers, stray-light)
- Use of well-known proteins for comparison of results
- Scientific excellence, with peer review evaluation

- Multi-national research teams
- Open communication, publications
- Long period from experiment definition until mission
- Public funding.

A number of facilities were developed for this type of research. In Europe, ESA's APCF, the Advanced Protein Crystallisation Facility, is the facility in which this type of research is done. It was also successfully used by American scientists, who had access to this facility via a bilateral agreement between ESA and NASA. The APCF has been flown five times so far and is planned to be flown again about once a year until the Space Station is available, where the APCF will be the first European payload. It offers video observation and all standard growth methods, allows customisation of growth reactors to specific needs, and offers an interferometer to detect early crystallisation and investigate the growth zones.

Another instrument for the Space Station is currently under development; this offers additional sophisticated diagnostic tools, such as a Mach-Zehnder interferometer with phase shift technology, and a light scattering device.

2.4 Competitive Commercial Research

The competitive field is quite different. The main emphasis is on the structural analysis of specific target proteins, which are of importance for the industrial research group of one company, normally an "innovative" protein. In such cases, confidentiality plays an important role, peer reviews are not accepted (and also are not helpful, as they use other criteria!), and analysis or diagnostic tools are not required. On the ground the research is privately funded and, also for space research, the industry effort is not supported by public money.

The competitive type of research is characterised by:

- Growth of large high-quality crystals (in space)
- Growth methods compliant with ground laboratory standards
- No diagnostics, but as many samples as possible
- Investigation of structurally unknown, hard-to-crystallise proteins
- Commercial value, market potential
- Single corporate research team
- Confidentiality, no publications, intellectual property rights
- Ad-hoc access to space in a few months
- Private funding, complemented by public funding.

As a consequence, the space research facilities look different; they have no diagnostics, the main goal being to accommodate as many growth reactors as possible into a given volume to allow the maximum number of experiments.

However, not all such designs are successful. Some facilities had too tiny reactors for really large crystals. Low-cost facilities could provide lower access costs, but the experimental results may then be poorer, as the advantages of microgravity are compensated by the disadvantages of the lower quality of equipment.

3. The Commercial Protein Crystal Growth Facility (CPCF)

Figure 1. The CPCF shown here has dimensions 400mm by mm by 100mm by 93mm. It contains 20 APCF-type crystal growth reactors, which are activated/deactivated by turning the front-mounted knob.

The CPCF is a 4 kg derivative of ESA's 26 kg APCF (Advanced Protein Crystal Growth Facility) and uses the same type of reactors which have been described in detail [Reference 1]. The secret of this successful APCF reaction chamber relies on many demanding design and manufacturing details. A modified version for use inside the ISS is under construction.

The small CPCF (Figure 1) does not provide any thermal control, which means that it has to be accommodated inside a thermal enclosure. As many space proven incubators exist and fly on every mission, it is possible and easy to fly the CPCF "piggy-back" inside such an incubator. As the CPCF also does not include any electronics, the interfaces to the spacecraft are very simple. The facility will be filled on the ground, inserted in the incubator, launched and then — in orbit — started (and later stopped) by manually turning the knob (clearly seen in Figure 1).

The CPCF has 20 experiment reactor sites, suitable for each of the standard growth methods and individually adaptable over a certain range to specific protein volume needs:

- Hanging drop (i.e. vapour diffusion)
- Free interface diffusion
- Dialysis.

The CPCF provides double containment of the experiment fluids to meet NASA's safety requirements.

4. The First CPCF Mission

The mission STS-95 in October and November 1998 was the first mission with the CPCF. The project was started in May 1998, with the design and manufacturing of the hardware, and the customers from German pharmaceutical companies were approached at the same time.

The hardware was developed under DASA's responsibility and mainly with its own funds. The CPCF mission organisation (flight procurement, negotiations with NASA safety) was under DASA's responsibility as well, but the mission-related costs were financed by DLR. In this way this project was a pilot project to test the feasibility of a new approach, giving space industry the full responsibility in order to respect the requirements of commercial customers.

The industrial partner in the US for this mission was the company ITA, in Philadelphia, which organised the Space Shuttle mission, provided the incubator and took over the interface to NASA. This cooperation was very effective and led to good results. However, it would also be possible to use other flight opportunities, such as Spacehab or the Russian part of the ISS.

The mission schedule was as follows:

- 22nd October: support filling of reactors at customers' premises
- 22nd October: transport to launch site in thermally controlled environment
- 22nd October: finalise experiment material verification
- 29th October: launch of STS-95
- 29th October: activation of CPCF
- PROCESSING OF CRYSTAL GROWTH FOR 8 DAYS
- 6th November: deactivation of experiments
- 7th November: landing at KSC
- 8th November: retrieval of CPCF
- 12th November: transport to Europe in thermally controlled environment
- 12th November: distribution of reactors to customers' premises.

The mission was highly successful from technical and organisational points of view. It has been demonstrated that a mission can be organised on a commercial/industrial basis, respecting the main requirements of the user industry.

5. Lessons Learnt

Some major aspects of the CPCF mission which are briefly discussed include:

- Fast access and *ad hoc* participation
- Success
- Confidentiality
- Cost effectiveness
- Effort for the customers.

5.1 Fast Access and Ad hoc Participation

The development for this mission was realised in less than six months. One of the six experimenters changed his protein three months before the mission; two experimenters changed their experimental protocol and chemicals used. This shows the typical requirements of industrial users, which have to be respected.

Ad hoc participation means that the program has to be started without firm commitments from the customers. Waiting at the initiation of a programme until there is a sufficient number of firm commitments from the customers is not realistic. All customers have indicated that they would continue using the CPCF.

5.2 Success

The mission was a success regarding the organisational, operational and technical aspects. The harvest of crystals showed small crystals only. According to current investigations this might be due mainly to the short mission duration. Further evaluations of the activities are being performed.

5.3 Confidentiality

The customers had far reaching requirements on confidentiality:

- Substance to be crystallised had to remain confidential
- Experiment protocol
- Composition of chemical compounds
- Name of pharmaceutical company.

On the other hand, NASA has to guarantee mission safety, and the government has to respect the justification issue of public funding.

In the CPCF mission the following approach was realised to cope with these conditions: DASA took the role of Principal Investigator, not disclosing the names of the participating industrial companies but answering to DLR for the soundness of the industrial experiments.

Details of experiment protocols limited to chemical and toxicological aspects were related in a way fully meeting NASA's respected requirements, but not revealing any proprietary information.

5.4 Cost Effectiveness

As the facility was developed as the derivative of an existing design, the development costs were low. The mission costs were also minimised, as the CPCF does not ask for any resources except thermal enclosure. The mass and hence the corresponding launch ticket prices are low. The CPCF will be further improved by modification of the reactor geometry, which will allow a higher number of reactors in the same volume and mass limits. The existing design also kept the interface costs to NASA down to a reasonable level.

5.5   Effort for the Customers

The approach for space experiments necessarily differs from everyday screening ground trials. Some special proteins of high commercial importance might even, in an industrial laboratory, be the subject of a more sophisticated growth trial procedure. Nevertheless, the special efforts for space experiments should be as effective as possible to make μg-experimentation become a "standard-industrial" tool. The CPCF pilot project took a path to reach this major goal, keeping the typical "space work" (e.g. paperwork, special testing) to a minimum acceptable to the industrial customer.

It was not necessary for our customers to move from their laboratory. The customers received laboratory models of our reactors for ground preparation and filled the flight units in their laboratories a few days prior to the launch. They received the material shortly after landing. Such an approach is very attractive, especially for workers in industrial laboratories.

6. The Malaria Experiment

More than 200 million people are suffering from malaria worldwide and more than two million persons, mainly children, die every year from this disease. Available treatments are beyond the reach of Third World countries and of low effectiveness because of the poor general physical condition of the population in these countries. A medicine derived from a known substance and specifically developed for this parasite could be a most worthwhile remedy.

In addition to the confidential projects of our customers, we were also involved with a university group with similar objectives to those of our industrial customers. The malaria experiment is a project of public nature, and the results can be distributed to the public. Protein crystal growth experiments are performed on one enzyme of the malaria parasite *Plasmodium falciparum*. If structural analysis of this enzyme or enzyme pharmacon complex is possible using one of the crystals grown, there will be a good chance of developing an effective medicine for curing malaria. This drug will be a modification of the chemical substance methylene blue.

References
1.   Bosch, R. et al.:*Journal of Crystal Growth, Vol.* 122, pp. 310-316, 1992

The ISS, an Opportunity for Technology in Space

J.Tailhades, Matra Marconi Space, 31 Av. des Cosmonautes, 31402 Toulouse Cedex 4, France

e-mail : jacques.tailhades@mms.tls.fr

D.Routier, Matra Marconi Space

e-mail : daniel.routier@mms.tls.fr

Abstract
Within 5 years, the International Space Station will become a permanently inhabited space system, which will provide an easy access to the Space Environment.
Up to now, the early utilisation of the ISS has been planned for different classes of activities mainly led by scientific objectives. Due to ISS utilisation costs, a significant part of available resources will be made accessible to non-scientific and non institutional users, thus allowing to shape a real commercial utilisation of the Space Station.
This commercial utilisation will address Space industry as well as non-Space industry for various types of utilization which may be classified as :
applications using the ISS external environment (thermal, radiative, vacuum, ...)
micro gravity applications,
media applications.
MMS, deeply involved in the development of the COF (European module of the ISS), has started, beside the scientific facility (BIOLAB, MELFI,...) development and the integration of External Payload on the ISS truss, commercialisation study of these facilities. The objective is really to identify the possible markets through non-institutional users, which are ready to invest in Space utilisation for their own needs.
Then, the scope of the paper will be to present a preliminary review of the commercial applications of the ISS available to Space industry and what conditions have to be made available for enabling the development of such market.
It will be completed by a review of non-space applications and the identification of possible supports to be offered by Space Industry to non-space one for accessing the ISS.
Through several examples, the paper shows how ISS derived markets may be shaped. The focus is on typical technological applications and their benefits to industrial users.

1. Introduction

Within a few years, the International Space Station will be available for science and technology as a permanent laboratory providing full access to the Space environment through internal facilities and external payload accommodation sites.

In the same time, reducing costs and improving the performances of future missions will require permanent engineering research and technology development to develop new classes of products for Space applications, which may also benefit to non-space applications.

41

G. Haskell and M. Rycroft (eds.), International Space Station, 41-49.
© 2000 *Kluwer Academic Publishers.*

Up to now, Space technologies are always tested and validated on-earth using complex test facilities and modelisation of Space environment constraints to contain as much as possible the inherent risk due to later Space utilisation.

Using the ISS as a testbed for improving technology will be a key issue of the future technology development programme which needs to be evaluated now for preparing its real utilisation.

2. The ISS, a tool for Space Application

The ISS has been developed to provide a permanent laboratory in Space allowing to develop various scientific programs all along its twelve years lifetime. In a preliminary approach, fundamental science has been targeted through the definition of the Microgravity facilities for Columbus located in the pressurised area of the ISS:

- the BIOLAB, a laboratory for life sciences and micro-biology
- the MSL, a laboratory for Material sciences
- the FSL, a laboratory for Fluid Sciences.

Besides these large facilities, the ISS provides a set of generic devices :

- the European Drawer Rack (ESA)
- the Express Rack (NASA)
- the Cold Laboratory Support equipment in charge of providing cold storage area to scientific samples : MELFI (up to –80°C), CRYOSYSTEM (80K), Refrigerator Freezer (-20°C).

Naturally the NASA and NASDA modules are also filled with large scientific facilities providing tools for science in Space.

In the same way, the ISS truss is equipped with external adapters allowing to accommodate on Express Pallet Adapter scientific payloads to benefit from the Space environment outside the ISS. These ExPA are located both on the Earth and Deep Space directions providing a wide variety of places to accommodate P/Ls. The Columbus module as the Japanese one has been fitted with locations to accommodate Express Pallet. In the current definition of the Express Pallet programme, a multi-user facility, EuTEF (European Technology Exposure Facility), is presently under development by ESA to provide any class of user with a cheap and performant access to Space environment for technology reasearch in any domain.

Still under investigation, Free Flyer Micro Operators have been defined as being a third class of devices. Such micro satellite are able to accommodate payloads in the vicinity of the ISS which allows to benefit from its servicing capabilities and to offer a high level of microgravity and a full independence to the payload during its utilisation in a free flying mode.

Then, in a synthetic way, ISS can provide three types of Space access and resources :

- internal microgravity payloads
- external Express Pallet
- autonomous Free Flyer providing the same interface capability as the ExPA.

From a schedule point of view, the two first classes will be made available as soon as the ISS will be assembled (about 2004) while the third one will be available in the second part of the ISS utilisation (about 2007).

3. Technology in Space

3.1 Classification

Technology in Space may be classified according to its utilisation :

- <u>low level technology</u>, mainly covering the basic components which will be used in equipment. Such component address as well mechanical, as thermal, as electronic parts which may be necessary to be demonstrated in Space before any utilisation. As a simple example, one can think to the EEE parts of which tolerance to the Single Event upset, and capacity towards cumulated ray dose have to be established and demonstrated prior any utilisation. New material providing shape memory need to be demonstrated before their selection for Space project and always require complex test facilities
- <u>medium level technology</u>, mainly addressing Space Equipment. They cover the full range of application used in all the different parts of a Space System as the sensors, the processor, payload devices or support equipment as the fluid loop for thermal devices. In most of the cases earth proofs are sufficient to qualify them, however in some cases the Space environment cannot be modelled on-ground and complex test devices, including parabolic flights are necessary with inherent limitations tas the microgravity duration
- <u>High level technology</u>, mainl addressing new instrument or system component which have to be agregated to provide the Space system with

some functions. As an example such kind of technology may be considered in the development of new classes of Instruments as the new programme of Space Lidar which cannot be fully tested on earth with existing facilities.

3.2   What conditions for validating the Technologies?

In any case mechanical, thermal, electrical parts have to be verified requiring various tests facilities:

- thermal environment, aiming at simulating solar flow under vacuum conditions
- thermo-mechanical environment, aiming at simulating the constraints generated by the permanent modification of lighting conditions in Space according to the motion of the carrier
- radiative environment, aiming at modelling the impact of particles on components (materials, EEE parts, ...) exposed to Space flow
- deep space vacuum, aiming at studying the impact of off-gassing on structures and the impact on their physical properties (stability, ageing, ...)
- ...

It remains obvious that Space Industry is now able to set up test facilities for most of the technology validation needs. However, the main constraints come from the definition and preparation of adequate facilities, the duration of the tests to be run in order to obtain representative and reproducible results, the cost of such test devices and the sharing of these tests devices with several users to reduce the utilisation cost.

Furthermore, the Space market is becoming more and more commercial which imposes to reduce the technology qualification time in order to increase the competitive advantage when owning a particular technology.

3.3   Who are the users?

The analysis of potential needs in technology shows two different classes:

- Space Industry, comprising as well companies developing equipment and systems as companies using Space systems. In any case the development and the utilisation of technology, which may improve their market share and the associated profit, interest them
- Non Space industry, developing products and systems for earth applications which may be interested by accessing Space environment to

benefit from representative and severe environmental constraints allowing to assess the applicability and the performances of the tested technologies.

In the current Space programme development, new technologies appear which cannot be fully tested and demonstrated on-ground and are requiring innovative approaches to be set up in the future. For example fluid loops are under investigations to solve the thermal control of spacecrafts by providing easy to accommodate solutions. Unfortunately their behaviour is fully dependent on gravity conditions and results obtained on-ground cannot be translated as such under Space conditions. Effective in-orbit demonstration seems to be the only acceptable proof-of-concept.

4. Why ISS may be Considered as an Opportunity

4.1 ISS opportunity

Reducing time-to-results is directly linked to the competitiveness of new technologies to increase the return on investment of investing company.

Then, ISS is a real opportunity because it provides users with a permanent, easy to access and easy to retrieve Space environment:

- permanent : during its 12 years lifetime, the ISS will be regularly services by the shuttle and the ATV allowing a regular payloads upgrade and/or changes on-board the ISS.
- Tools designed for users : a set of equipments, facilities have been defined in the ISS program on NASA and ESA sides to ease the ISS utilisation :

  - Development of SPOE devices which may be used as building blocks of Space experiments devices,
  - Development of EuTEF providing a technology test environment facility for exposing "devices" to Space environment in a fully controlled way,
  - Express Pallet Adapter, providing a interface plate and associated system components to install payloads outside the ISS,
  - Multiuser facilities providing users with standardised containers for accommodating test devices under controlled and adapted tests sequences.

- Turn over ISS is the unique Space System which guarantees long time Space conditions and capability to retrieve on-ground the payloads. This means that it becomes possible to prepare an experimental device for flying on-board the ISS, to use it on-board during several months and then to retrieve on-ground for analysing the Space environment impact.

As such, the ISS combines all the test environment needs plus the capability to access an existing command/control environment which allows the user to focus on the technology accommodation and not on the surrounding components required to operate it.

The ISS is in competing:

•   Small satellites, which offer an access to Space environment with no capability to retrieve it. However fully dedicated to the device under tests, they allow to simulate in a complete way the working conditions
•   Ground facilities, which allows limited simulations of Space environment but provide an easy access and easy utilisation.

The market is moving from the utilisation of Ground facilities to the utilisation of Small satellites or Shuttle (SpaceHab) on a case-to-case basis, which has to be reviewed with the hypothesis of ISS utilisation.

4.2 ISS needs

To be used by industry as a opportunity for Technology, the following issues have to be fixed in the coming years to support private users:

•   Time to results, i.e. the duration between the moment where a company has a need for Space demonstration and the moment the results become available. Up to now this time is quite long due to the limited number of opportunities, the selection methods and it is really a major issue to guarantee the time to get results in order to get the ISS competitivity versus the other methods
•   Confidentiality, which covers as well the privacy of the produced data in order to remain only known by the sponsoring user as the identity of the sponsor which, in most of the case, does not want to be known in order to get a competitive advantage towards its competitor on the commercial market. In particular, when a company is aiming at demonstrating a new technology, it may be of the utmost importance to maintain it unknown to avoid disseminating its own ideas
•   Intellectual property rights: the ISS is an international area governed by a set of MOU between the different partners. Then, it is not known what is the applicable law to each module of the ISS according to its owner. This problem needs to be clarified in order to get the right confidence to the users in the fact that the results are fully protected and, by the way, the intellectual property is guaranteed in Space as on Earth
•   Allocation of resources: as the ISS is providing a limited set of resources, any user needs to really know in advance what level of resources will be

made available to its experiment in order to manage correctly its technical and scientific plans whatever the ISS utilisation conditions are. This means, in other terms, that the planning of the ISS resources profile allocation should have to be prepared and agreed with the user and maintained on-board

- Definition of a charging policy, i.e. the precise definition of the ISS utilisation cost allocation to the user. In fact, using the ISS requires taking into account the different components of a mission: launch cost, operation preparation and execution cost on-board resources utilisation costs... In order to give confidence to potential users of the Space station the Agency has to clearly propose a cost policy according to the level of funding share between the Agency and the users. The Spacehab example is, today, a first attempt to reach that cost transparency, which is the only way, to ensure the success of the ISS utilisation by non funded users. An identified approach consisting in allocating launch cost to Agency and Payload definition one to the user is not fully satisfactory because it does not take into account the wide variety of potential applications and potential sharing between the users and the Agency

- Selection: the selection of the payloads to be flown is handled by a pier group at ESA gathering scientific objectives as well as programmatic ones which need to be reviewed in order to establish a new approach for selecting the flight rights of the industrial users. This choice will have to take into account the interest of the Space utilisation as well as the potential benefit for business. This issue is a challenging one to guarantee the access to the ISS under competitive and commercial conditions

5. An Example of Space Industry Role for ISS Utilisation

In the ISS field, MMS besides its important contribution to ATV and COF programme is responsible for :

- the development of the MELFI, -80°C freezer of the ISS to be considered as a support for biological research
- the development of the BIOLAB
- the integration of the Express Pallet Adapter under responsibility of ESA to be located on the ISS truss and later on the COLUMBUS node.

By these projects, MMS gained the complete knowledge of the ISS access and rules for Payloads design, development and accommodation, as well a consistent involvement in the ISS utilisation plan through MAP projects.

Today, it is important to define and work on the conditions, which will allow industry to use the ISS as a simple test facility for its own technology.

Although, in the future it will be required to propose to users set of tools and devices making easy the access to the ISS.

For instance the EPI programme has defined a generic system layer which provides any grouping integrator (of instruments on an Express Pallet) all the interfaces with the ISS. This first step will allow to propose to user (space and non space industry) services for accommodating their P/Ls if required in a fully confidential way although fulfilling all the ESA/NASA safety constraints.

In the same way, MMS and the other prime contractors of the MFC are initiating a promotion activity to support the utilisation of internal Columbus facilities to private users. This effort needs to be a very long and continuous one to get ready by 2004 but is mandatory to benefit from the ISS opportunity. MMS as such is also interested in the utilisation of ISS as technology testbed which may be accessed for:

- verification of thermal behaviour of fluid loops under microgravity
- verification of a new concept of cryogenic link between a cooler and a dewar under microgravity conditions
- verification of materials ageing under Space environment conditions
- assessment of Space components (EEE parts, ..).

This approach is emerging for the next ten years with the reality of the ISS which may be used for such kind of activity. Up to now the selection of Space demonstration was slowed by the lack of opportunities and the difficulty to be phased with these opportunities which has also to be improved in the ISS utilisation planning to establish a effective commitment between users and ISS responsible.

6. Conclusion

Using the ISS for improving technologies will be a key of the success of the ISS by installing a permanent tool to the service of technological research.

The applications field is very large and rely on existing facility developed for such purpose as the EuTEF, but also on generic approach (ExPA, FFMO, MFC, ...) which will be accessible to users as soon as Space company as MMS will offer the services for preparing, installing and verifying the experimental device before the flight.

Then using ISS in such way will create new markets:

- technology accommodation

- exploitation and operations
- support to mission preparation.
- ...

This commercial approach will become only if ISS access rules are applied to that objective providing industry with all the necessary guarantees and if companies are preparing through early opportunities missions the necessary services and support devices to be offered later on.

ISS: Management of the Commercial Research and Development Process

R. Moslener, The Boeing Company, MC HS-32, 2100 Space Park Drive, Houston, TX 77058, USA

e-mail: ralph.n.moslener@boeing.com

Abstract
The paper discusses ways in which the many different experimental facilities aboard the ISS can be managed and used in a commercial context which is both customer-friendly and service-orientated.

1. Introduction

The International Space Station (ISS) is a space laboratory, whose purpose is the expansion of knowledge to benefit all people of all nations. When completed in 2004, it will be vast, as illustrated in the central panel of Figure 1. Surrounding this panel are images of various modules in the construction phase and, at the bottom, pictures of the Integrated Equipment Assembly (left) and the photovoltaic array, or solar cells (right).

Figure 1. The International Space Station, and some of its component parts

G. Haskell and M. Rycroft (eds.), International Space Station, 51-57.

The capabilities of the ISS are numerous. As shown in Figure 2, the research planned for the US laboratory (shown in the center) ranges from life science to space science, microgravity science and remote sensing for Earth science, plus engineering research and technological studies, and investigations leading to developing new products in space.

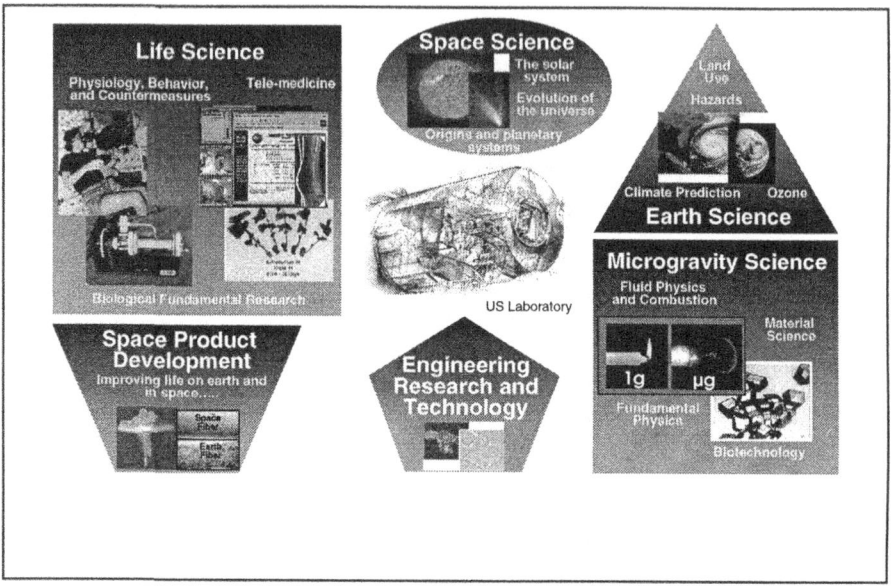

Figure 2. Diagram illustrating the many and varied capabilities of the ISS for research

General market surveys indicate that scientific investment in ISS utilization will start slowly and grow over time (Figure 3). As the ISS capability grows, industrial acceptance of the comparative advantage of using the Space Station should grow proportionally. On the other hand, interest in and the value of sponsorships will start quickly and diminish over time unless demand curves shift in relationship to successful scientific utilization.

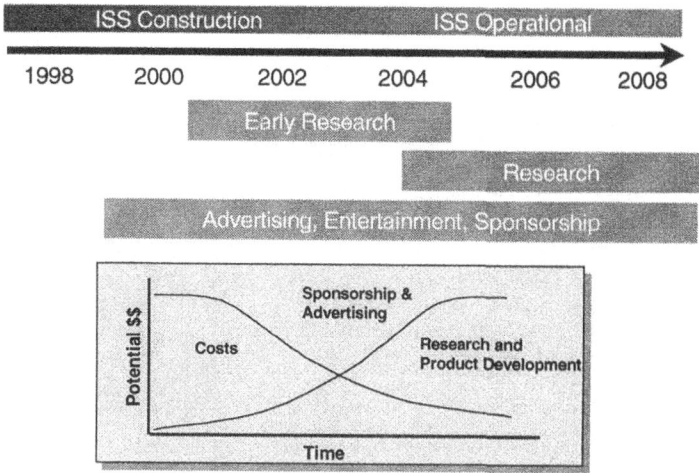

Figure 3. Ten year plan for the ISS

2. Details

In order to carry out research and conduct investigations, a number of general purpose facilities are planned aboard the ISS. In many of these, the effects due to the microgravity environment in an Earth orbiting vehicle are explored. For example, Figure 4 shows the fundamental purposes of studies of biotechnology processes in the Biotechnology Facility (shown on the right).

Description/Purpose
The basic goals of the Biotechnology Science Plan are to: • Understand the fundamental roles of gravity and the space environment in biotechnology processes   • Use low gravity experiments for insight into the physical and biochemical behavior of biotechnology processes   • Apply the scientific knowledge needed to analyze, quantify and improve these processes   • Contribute to Earth-based biotechnologies that enhance our health and quality of life   • Develop biotechnology to specifically support space exploration

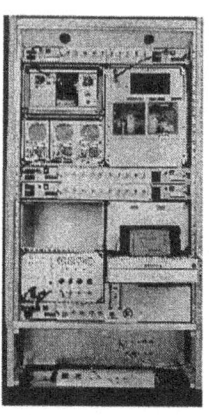

Figure 4: Equipment for, and goals of, biotechnology research

Similarly, Figure 5 shows a diagram of the ISS equipment for research on a range of biological specimens. A centrifuge will be available to produce a gravitational acceleration up to 2g, twice the value experienced at the Earth's surface.

Description/Purpose

• Supports the Gravitation Biology & Ecology Research Program on Space Station

• Promotes understanding of the role and influence of gravity on non-human living systems

• This facility will be used to determine the effects of the space environment (e.g., from radiation and microgravity), on biological specimens from cells and tissues to whole plants and rodents

• The centrifuge rotor will produce a gravity environment from 0 to 2 g's. These facilities will support the study of the levels and duration of gravitational force necessary to offset the deleterious effect of microgravity

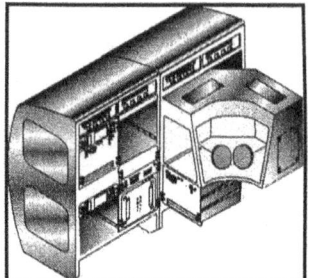

Figure 5. Diagram of equipment aboard the ISS for research in biology, and the purposes of such investigations

Figure 6 illustrates two different racks of equipment to study the behavior of fluids in the microgravity environment (left), and combustion phenomena under microgravity conditions (right).

Description/Purpose

• Supports Human Exploration and Development of Space (HEDS) objectives by facilitating sustained, systematic microgravity fluid physics and microgravity combustion science

• Will be a permanent multi-discipline research facility occupying three payload racks in the United States Laboratory Module

• The on-orbit facility consists of a Combustion Integrated Rack (CIR), a Fluids Integrated Rack (FIR) and a Shared Accommodations Rack (SAR), supporting 12 typical microgravity experiments each year throughout its 10-year design life

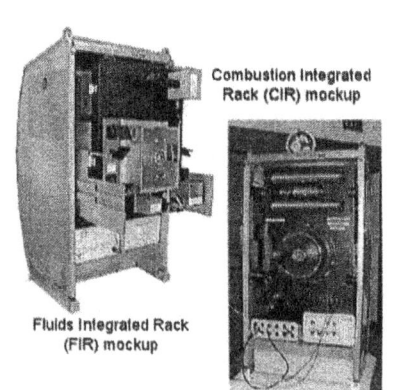

Combustion Integrated Rack (CIR) mockup

Fluids Integrated Rack (FIR) mockup

Figure 6. Mockup of equipment racks for studies of fluids and of combustion under the microgravity conditions experienced in Earth orbit

To stimulate new, small experiments aboard the ISS, a general purpose rack to Expedite the Processing of Experiments to the Space Station (EXPRESS) has been developed. It is shown in Figure 7, where its features are described briefly. The Active Rack Isolation system mentioned is a system to improve the microgravity environment which is disturbed by the activities of the astronauts. It is a complex mounting system which can be operated in all six degrees of freedom simultaneously.

Description/Purpose

- The EXPRESS Rack concept was developed to support small payloads on orbit with a shortened ground integration period

- It accommodates multiple payload disciplines and supports the simultaneous and independent operation of multiple payloads within the rack

- It is launched with initial payload complement and remains on-orbit allowing payloads to be changed out as required. Four of eight EXPRESS racks will include the Active Rack Isolation System

EXPRESS Flight Rack with secondary structural members installed

Figure 7. The EXPRESS rack

Another facility developed with many different users in mind is the Microgravity Science Glovebox (MSG) shown in Figure 8. Small scale experiments on effects due to microgravity can be carried out with chemical and biological containment by ISS crew members.

Description/Purpose	
• A multi-user facility that enables users to conduct small science and technology investigations. An enclosed work volume which provides resources necessary to conduct a wide range of microgravity research • Investigations are expected to include: 　• Fluid Physics 　• Combustion Science 　• Materials Science 　• Biotechnology 　　(Cell Culturing and Protein Crystal Growth) 　• Space Processing 　• Fundamental Physics 　• Technology Demonstrations	

Figure 8. The Microgravity Science Glovebox

There are other facilities aboard the US laboratory on the ISS, such as the Materials Science Research Facility, a facility based on the EXPRESS rack for mounting instruments to view through a high quality window, and a facility for evaluating the physiological, behavioral and chemical changes which occur in astronauts during their space travel.

3. Discussion

Use of these facilities by the government, academia, and industries requires cooperative research and development to utilize the ISS fully. A better appreciation of how space-based markets will open up will lower associated risks and give a return on investments. It will take some years for new commercial markets to be developed and fully utilized. This is not only because of the time needed to construct the ISS to its complete specification but also because of the challenges of creating new businesses when the costs differ greatly for those currently being experienced in the industry. However, it is anticipated that the attraction of the dramatic possibilities for solving problems aboard the ISS will accelerate its acceptance by industries around the world who wish to gain a competitive edge. As is always so, they will need to make compelling cases to invest in costly studies such as those to be performed aboard the ISS.

Currently the Boeing organization worldwide is actively involved in the ISS program — its implementation, operation, and the reliable support of both payloads and products. Aspects of these can be combined to bring substantial benefits to commercial users. Boeing is aiming to provide a customer-friendly, "service-oriented" environment to the users, which is on-line and interactive, thereby facilitating users of the ISS. Payload, product and data security and integrity are essential from the users' viewpoint, and Boeing will assure these. Boeing will reach out to potential users, and educate them in the capabilities of the ISS and the available resources. But Boeing and the users must each have well defined roles and responsibilities, with detailed costs and adhered to schedules.

4. Summary

Managing the ISS commercial research and development process requires Boeing to understand:

- ISS capabilities for research
- market developments
- potential sources of commercial support
- implementation models of ISS utilization
- the needs of the users.

Thus Boeing's role is a complex one — as facilitator, integrator and value-added partner for many users of the ISS

Report on Panel Discussion 1:

Management of the Research & Development Process

O. Gurtuna, C. Rousseau, International Space University, Strasbourg Central Campus, Parc d'Innovation, Boulevard Gonthier d'Andernach, 67400 Illkirch-Graffenstaden, France

e-mail: gurtuna@mss.isunet.edu, rousseau@mss.isunet.edu

Panel Chair : M. Uhran, NASA, USA

Panel Members:

O. Atkov, Cosmonaut, International Space University
K. Knott, ESA
T. Kuroda, NEC Corporation, Japan
R. Moslener, The Boeing Company, USA
J. Vaz, BRAZSAT, Brazil

The panel discussion, involving participants from both the public and private sectors, focused on the needs of the private sector for using ISS as an R&D facility and the role of the government for providing the necessary infrastructure for its commercial utilisation.

During his introductory remarks, **M. Uhran** asserted that the pricing policy and the schedule of access, together with the competitive advantage of using the ISS for commercial purposes, should all be well defined before one can talk about creating a space marketplace. **J. Vaz** commented on the importance of getting the know-how and personnel training for developing nations; he stated that for such countries buying access to space does not necessarily enable the transfer of crucial expertise.

Some of the speakers cited promoting ISS utilisation to the non-space sector as one of the responsibilities of the sellers in such a marketplace. This represents a shift in the strategy of the sellers, which, up to now, had focused primarily on the space sector as users of the ISS.

O. Atkov presented two important lessons that were learned during the Mir program. Good advertisement of the capabilities of the Space Station and exhaustive ground planning and testing for R&D projects before operating them in space are two keys to success.

G. Haskell and M. Rycroft (eds.), International Space Station, 59-60.
© 2000 *Kluwer Academic Publishers.*

The need for potential users of the ISS to have better information was raised by some of the participants. It was argued that explaining the pros and cons of the Space Station to the users was an essential step. The users need to be informed even if the content of the information might be negative, as in the case of the microgravity environment being adversely affected by crew activity.

To define the unique features and the core competencies of the ISS was identified as another essential step. **R. Moslener** stated that the ISS crew, from several nations, will be the most precious asset in orbit. It was agreed that studying the long-term effects of microgravity on human physiology will be an important aspect of utilisation of the ISS.

Some participants emphasised the importance of a sound regulatory framework for commercialisation. Protection of intellectual property rights and the provision of tax incentives for commercial users were the main topics of this discussion. The "zero gravity, zero tax" approach proposed by The Honorable Dana Rohrabacher, Chairman of the U.S. House of Representatives Space and Aeronautics Subcommittee, in his live video address to the Symposium, was examined. The general agreement was that any public initiative to support ISS commercialisation will help the private sector.

Session 2

Entrepreneurial Initiatives to Use ISS for Profit

Session Chair:

J. K. von der Lippe, INTOSPACE GmbH, Germany

Entrepreneurial Initiatives to Use the ISS for Profit

J. K. von der Lippe, INTOSPACE GmbH, Sophienstr. 6, D-30159 Hannover, Germany

e-mail: lippe@intospace.de

Abstract
In the previous years of space station development the focus for utilization has been on research under Space Conditions, primarily basic research. Considerations for commercial utilization were mostly regarded as getting the non-space industry to use the station for application-oriented research or even manufacturing. Now as the ISS has made its first step of assembly the plans being drawn up for ISS utilization will acknowledge even business beyond any yet accepted commercial activity in manned space systems of the Western world. This is a great chance to make real progress in the commercialization of manned systems which has been discussed for a long time. Future commercial initiatives will have to reflect the specialties of the ISS, which are:

- It is a manned system with an exceptional international character
- It is a highly sophisticated research centre
- It will have very high public attention every day
- It is a low Earth orbiting platform with ample resources and regular manned access
- It has the capability to grow.

This paper reviews the various potential opportunities to use the exceptional conditions of the ISS which make it an ideal place for new business opportunities beyond those presently planned. These will be in the domain of public services (entertainment, advertisement, product placement, tourism, etc.) which will attract new entrepreneurs, without being in conflict with the research centre core activities.

1. Introduction

Through all the years of Space Station development the focus for utilization has been on research under space conditions, primarily basic research making use of microgravity conditions, the unique feature provided by the ISS and not available on Earth to that extent. The aspect of commercial utilization, however, became a topic at periodic intervals whenever the budget discussions required that. The first commercial initiatives concentrated on using the results of basic research for application on the ground, performing application-oriented research with industrial involvement or even considering the production of high value material in space. Now, as the International Space Station has begun its assembly the perspectives for commercial activities are expanding and business so far unthinkable (because of governmental rules) are becoming part of the ongoing planning of its utilization.

This is a great chance to make progress in commercial business using the Space Station, even in areas beyond the classic fields such as applied research and technology development for industry.

G. Haskell and M. Rycroft (eds.), International Space Station, 63-68.
© 2000 *Kluwer Academic Publishers.*

2. Commercial Industrial Utilization Fields

The research programmes sponsored by space agencies around the world have provided enough information to judge the value-adding potential of this new research tool for application-oriented research for use by industry.

The utilization of the space environment, particularly the microgravity condition, has developed to be a research tool used for the optimization of processing and improvement of products on Earth, and so providing benefits to industry. The International Space Station with its capability for regular access and continuous utilization potential provides for the first time an ideal test bed for industry. The fast growing commercial space market for communications, navigation and Earth observation requires industry to enter into an increasing competition with newly developed satellite technology. The ISS will provide the involved space industry with the opportunity to test and qualify their advanced technology in order to gain better market chances. The availability of the ISS for European industry provides a competitive advantage in fields of advanced technologies.

Besides the utilization of the ISS for industry, a considerable commercial business can be developed by industry providing operational and logistic services, either for government or private customers.

The permanent human presence in space requires a continuous supply of food, critical maintenance systems, and scientific equipment and samples. These demands are expected to produce an annual requirement of more than 50,000 kg of cargo. The performance of the related service to provide this logistic resupply depends considerably on the utilization of transport means such as foreseen with the NASA Space Shuttle system, the Russian Soyuz/Progress transporter or the European Automatic Transport Vehicle (ATV). However, the preparation and organization of this logistic effort and installation into containers will be an extensive operation which can be provided by experienced private companies.

This type of business has been demonstrated by the recent Mir missions, with the NASA shuttle system making use of the privately owned and operated SPACEHAB as a logistic container.

Further commercial business is expected to develop related to services required by users in the field of research and technology development with payload design and development, data transfers, etc..

Such potential industrial utilization — application-oriented research, technology development and testing, and the logistic support and related business — can be considered as the classical commercial and industrial opportunities. They are not the subject of this paper which discusses entrepreneurial initiatives.

3. Entrepreneurial Opportunities

In order to review the potential of the International Space Station for a business development, the uniqueness and specialties of the station have to be defined since they provide the possibility of added value.

The International Space Station (ISS) is:

- A manned system in low Earth orbit, with an exceptional international operation
- A highly sophisticated and professional research centre on the ground and in space
- A low Earth orbiting platform with ample resources and regular manned access
- Considerable potential for growth
- Very high public visibility.

The continuous high public attention of the ISS, influenced by the image and the admiration related to the profession of astronauts and the dreams of people derived from their vision of space travel to distant worlds, is an attractive environment for business in the field of promotion, public relations, advertisement, marketing, etc..

In summary it can be stated that the business platform which provides the basis for a creative entrepreneur is the international global outpost, a stepping stone to go to Mars, a scientific research centre of such high interest to the public that they never tire of reading about or seeing it on television and do not lose interest. Therefore it can be used in the following entrepreneurial examples.

3.1 Product Placement

Some commercial equipment will be regularly used by astronauts for the daily life or as tools or instruments for their work. These commercial products (e. g. laptop computers) can be qualified for space use and are therefore "space proved", and can be shown to the public during regular television transmissions or on special advertisements. Business can be developed in organising, promoting and implementing such product placements.

3.2 Education, Long Distance Learning

The ISS brings a new dimension to the education and know-how of the general public. The uniqueness of this global outpost can be used for education in schools, museums, public events, etc., through long distance learning, e. g. an astronaut teaching from the ISS via the Internet into selected classrooms. Commercial business is related to general management, sponsoring management, selling recordings and related educational material.

3.3 Advertisements

Since the International Space Station will command high public attention and any media event with the ISS will attract the general public, it can be used in various ways for the advertisement of products, either simple labels and products used on the station or as a secondary indirect utilization in print and video media. Again the profit making business is in development of innovative applications, marketing, coordination, management and — certainly — implementation.

3.4 Entertainment and Tourism

After several years into the programme of the ISS, with routine operations in research and industrial utilization, applications might become acceptable which are presently not considered realistic such as space tourism and entertainment.

The International Space Station, as a global research centre, can be compared to terrestrial large research centres, such as CERN or ESTEC, or the Kennedy Space Center, regarding its public attention and resulting in a considerable number of visitors. The assigned utilization of the ISS is therefore not necessarily in conflict with using the Space Station for public services. The concept for a first step in space tourism making use of the Space Station could be as follows:

- Target date: 2008

- Infrastructure: One living quarter (module) attached to the station.
                        One transport module (bus module)

- Transportation: Mixed cargo mode in the shuttle results in reduced
                        costs (MPLM, SPACEHAB, etc., + bus module)

- Clients: VIPs, rich people who dive today to the Titanic for US $
                        50,000

- Travel price: The US $ 10 million charged to a Japanese journalist for
                        a Mir mission needs to be reduced to less than US $ 1
                        million

- Entertainment: Experience microgravity, make observations of the
                        Earth from the cupola, have dinner with the
                        commander, visit the research facilities, etc..

The business potential behind these utilization possibilities of the ISS is seen to be in the addressing of mass markets, where revenues can be generated by huge numbers of customers, each paying a small fee, instead of by single customers, who have to invest a lot. The attraction of the ISS to the corresponding entertainment (advertisement and tourism) industry is expected to be in the representation of real space activities which mean a so far unreachable new product alternative. The provision of "real access" to a space-based human-tended infrastructure for the first time would be given not only to selected individuals but also to (almost) everybody. "Real access" in this context does not mean only tourism; it includes as well access via commercials or movie/TV productions.

While the topic "entertainment on the ISS" appears to space engineers and technicians to be somewhat far-fetched and not fitting in context with a publicly financed technical masterpiece like the ISS, it must be said that the purpose of the Space Station (besides being a scientific laboratory in space) is to promote and demonstrate international cooperation. And it is widely recognized today that successful international cooperation is based more on intercultural understanding than on the technical expertise of the partners. Now this strong link between technology and culture should be established with respect to the Space Station utilization as well. And it should not be limited to the cooperation between astronauts and space agencies; it should include the financiers, i.e. the tax payers, of the contributing partner states.

4. Prerequisites for Commercial Utilization

A successful commercial utilization of the International Space Station can only be built up in the areas briefly described above if the legal and economic environment is acceptable and business-friendly. Two basic cornerstones are essential:

- The general acceptance of the International Space Station as a world class research laboratory
- Implementation of the legal and commercial policies which allow non-scientific and non-technical business development on the ISS.

Most of the described potential commercial business has so far not been possible on the orbital infrastructure available in the Western world, e.g. Space Shuttle/Spacelab, and even the commercially operated SPACEHAB, because of legal restrictions imposed by NASA on the only available manned transport system, the Space Shuttle. The first steps in doing business with advertisements and product placements in orbit have been done on Mir.

A key issue of ISS commercial utilization for public services is the question of proprietary rights. That means who has the right:

- to allow a talk master to visit the ISS?
- to allow a TV station to broadcast directly from the ISS?
- to allow a company that would like to place its product with an astronaut?
- to receive money for allowing all this and selling these rights and the resources for transportation and utilization?

In order to enable the entertainment industry to include the ISS into their product lines and thereby to open new resources for financing the ISS operations, the corresponding framework conditions have to be established.

5. Conclusion

The business opportunities of the ISS as a global outpost are of very high potential for industrial users in the classical domain of industrial research and technology development. The short term profit making opportunities seem to be even more encouraging in the public service field; however, this business needs to be formally accepted as part of the ISS utilization programme.

From Space Station to Space Tourism: The Role of ISS in Public Access to Space

E. Dahlstrom, InternationalSpace.com, PO Box 60606, Washington, DC 20039, USA

e-mail: Eric.Dahlstrom@InternationalSpace.com

E. Paat-Dahlstrom, Space Adventures, PO Box 7584, Arlington, VA 22207, USA

e-mail: Emeline@SpaceAdventures.com

Abstract
Space stations have long been touted as the first step toward opening space for humanity. Visions of the future, such as in the 1968 movie "2001: A Space Odyssey," have shown space stations functioning as space hotels. And yet there always seemed to be a missing step between this vision of the future and the current space programs. How do we make the transition from the current government-funded space facilities, such as the International Space Station (ISS), to the private space hotels of the future?

There is a growing perception that the space tourism market could be large enough to drive the development of cheap access to space and private orbital facilities. Companies such as Space Adventures are already selling reservations for sub-orbital flights on vehicles under development. Large-scale space tourism could provide the market to support fleets of space vehicles offering cheap access to space, which in turn would benefit all other space activities. It is difficult to predict the schedule, but significant numbers of tourists going to orbit should be expected within the time of ISS operations. The ISS can play several roles to encourage space tourism, both technical and programmatic.

- Biomedical research aims should be conducted within the context of future public access to space, similar to the current context of Mars exploration. Research aims include improving space habitability, the medical implications of flying a broader spectrum of humanity (age, physiology, medical conditions, etc.), and on-orbit health monitoring and health care systems.

- The ISS can be used to evaluate and certify equipment or operations for future space hotels. Companies that develop ISS components should be able to apply their expertise toward the development of private space facilities, without undue restrictions nor unfair advantage. New companies, beyond existing contractors, should be encouraged to develop equipment for use on-orbit, initially for ISS and later for space hotels.

- ISS operations could be developed into an initial market for companies seeking to provide space equipment and services to future hotels and other facilities. This will require new roles for the government, and will threaten the monopoly of certain government offices.

Finally, the organizations that built ISS should not resist the next generation of space facilities. The builders of ISS have expressed the hope that ISS will eventually lead to the opening of space to humanity. When that time comes, they should accept that they have succeeded.

1. Introduction

The International Space Station (ISS) is now under construction, and plans are being made to encourage commercial activities on the station during its 15 year operational lifetime. Over the past 15 years of development, the ISS

G. Haskell and M. Rycroft (eds.), International Space Station, 69-78.
© 2000 *Kluwer Academic Publishers.*

program has dominated government human space programs. Commercial activities have transformed other areas of space, and human commercial space activities are now being considered.

The commercial space tourism industry exists today on a small scale, with dreams of large-scale activities in the future. Space tourism does not require the ISS – it needs low cost access to space. But, within the new promotion of commercial space station activities, some have proposed putting tourists on the ISS. The concept of such a tourist confronts all of the challenges faced by commercial ISS activities, and magnifies them with added symbolism. This paper does not propose placing tourists onboard ISS; the vision of space tourism is much larger than that.

The primary role of the ISS in public access to space will be in the transition from the traditional government space programs of the past decades to a future where space is an arena open to all the human activities which we accept on Earth. The ISS has, at least unofficially, been seen as a step toward enabling public access to space. But the traditional pressures of a government program ensure that public access will always be postponed. The ISS must be operated with the assumption that there *will* be a future for humanity in space. This paper proposes a variety of ways in which the ISS can enable the development of large-scale public access to space, and encourage the expansion of space industry.

2. Space Activities Today

2.1 Government and Commercial Space

Over the fifteen years of the Space Station program, much of the U.S. government infrastructure has remained the same, including the roles of the NASA Centers. There have been incremental improvements in the Shuttle's performance and reliability, but the system remains essentially the same. With the exception of Russia (due to the collapse of the Soviet Union), global government spending on space has not changed much in the last 15 years. The changes may seem significant to an industry so tightly coupled to government spending, but the scale of the changes has been modest compared to other industries. In contrast, commercial space activities have grown by large amounts over the last 15 years. Global commercial spending on space passed government space spending in 1996, and is growing at roughly 18 percent per year [Reference 1]. Launch vehicle contracts have gradually been shifting to government purchases of launch services. Commercial satellite production now dominates government satellite production. Today's global space industry has revenues of US $ 98 billion, and employs roughly a million people [Reference 1].

New communications satellite constellations have encouraged commercial ventures aimed at developing reusable launch vehicles (RLVs). The transformation toward commercial space activities has not yet affected human space flight, but this may soon change.

2.2 Space Tourism Today

Space tourism has the ultimate objective of placing tourists in space. Currently, we are at the beginning of space tourism, and the activities offered do not include space flights. Today there are a handful of space tourism companies, including Space Adventures [Reference 2]. Currently, Space Adventures offers customers tours, aircraft flights, and simulations as part of its "Steps to Space" program. Space Adventures is taking reservations for sub-orbital flights, and has agreements with several of the new RLV companies to arrange flights when passenger carrying vehicles become available. Currently, Space Adventures has roughly 70 people with reservations for US $ 90,000 sub-orbital flights. Other space tourism activities include parabolic ('zero-g') flights and high altitude flights in jet aircraft such as the MIG-25. Space Adventures, along with its partners in Russia, have flown roughly 150 people on zero-g flights (which currently cost about US $ 5,000), and thousands of flights on high performance jets (with prices ranging from US $ 3,000 to US $ 12,000 depending on the aircraft). Space Adventures is also beginning to offer tours to space-related sites around the world, similar to existing solar eclipse tours. Other evidence for the public's interest in space comes from the popularity of space mission simulations for children and the millions of visitors to space museums. A more direct measure of interest is the approximately US $ 2.5 billion earned by the movies in the "Star Wars" series, even without counting their spin-off products [Reference 3].

The development of RLVs for sub-orbital flights has been encouraged by the X-Prize Foundation, of St. Louis, Missouri. A prize is offered for the first company to build and fly a 3-person vehicle to 100 km altitude, and repeat the flight within two weeks. After such an experimental flight, the RLV company will need to develop a passenger carrying version of the vehicle, involving government certifications now being defined. Space Adventures has agreements with some of the sixteen X-Prize contestants. Space Adventures seeks to arrange flights for its list of customers, and these RLV companies want to demonstrate to investors that there exists a market for this service.

In recent months there have been renewed discussions of space hotel concepts and new entrants in the world of space tourism. Japan's Shimizu corporation has continued to showcase their vision of a future space hotel [Reference 4]. Hilton is now described as a partner in a concept using converted

Shuttle External Tanks [Reference 5]. Daimler Chrysler has announced plans for a 'Hotel Galactic' [Reference 6]. Robert Bigelow, of the Budget Suites hotel chain, has formed a division to design space hotels [Reference 7]. And Richard Branson has formed 'Virgin Galactic Airways' [Reference 6].

3. The Vision of Public Access to Space

What space activities do we want in the new century? What is our vision? Let us recall the vision presented in the 1968 movie "2001: A Space Odyssey" [Reference 8]. A commercial space plane docks at a future space station – designed as a transportation hub, with commercial restaurants and hotels. Flights to the Moon are as common as scheduled airline flights. Just from the perspective of space science, this low cost, routine access to space is a tremendous benefit. An essential part of that vision was the assumption of large-scale public access to space.

The involvement of the government transformed the public's view of space. World War II transformed research and development – tremendous projects could be achieved through centralized government sponsorship. It was, of course, the political and military competition of the Cold War that eventually took humanity into space. Ballistic missile capabilities were demonstrated with civilian government space programs, and the race to the Moon was used as a symbol of the conflict. In half a century we became accustomed to the government role in space. We have accepted the argument that it is such a difficult enterprise that only governments can make progress. But even on government space projects, private contractors perform the majority of the work. Within space, the example of communications satellites demonstrates what can be done after a transition from government research to private development. We need to step back from the world of space to see that this transition, from government to private, is the norm.

The history of technology presents many examples of difficult problems solved, at great expense, and then evolving into common consumer products (e.g., computers, television, video cameras, lasers, GPS, microwaves, the Internet). These products are first available for the wealthy, and then for the general public. In some cases, progress began with small, privately financed, research and development projects. In other cases, large government efforts developed the critical technologies. At some point, government control of these technologies has to be released, to spur new developments and innovations.

3.1 Global Tourism

What could support a future of large-scale public space travel? While the specific size of the space tourism market is difficult to quantify, the overall size of the global tourism market is very large – above US $3 trillion per year [Reference 9]. This number is growing, and may expand greatly with the addition of new tourists from China and other countries. The space community must think about serving markets on this scale, and not just the government space market. Consider the rate of spending of all tourists around the world – they will spend more than the annual budget for the ISS in just a few hours. During the course of this three-day conference, tourists around the world will spend more than the sum of all the annual government space budgets. Adventure tourists are now on the slopes of Mount Everest (despite a survival rate on the summit of only 80%), and thousands go to Antarctica each year. The hardships which these tourists endure far exceed what will be required of the early space tourists.

How much of this future tourism market might be spent on space tourism? In Japan, 70 percent have said they want to travel in space at a cost of three months' salary. In the U.S., surveys have found 42 percent would want to go [Reference 10]. These surveys indicate the pent-up demand for space tourism is on the order of US $ 1 trillion. Space transportation systems cannot yet offer a trip to orbit at the prices that would attract hundreds of millions of people. But we should expect space tourism to follow the pattern of other industries (cruise ships, air travel, etc.). Initially high prices would make the service available as a luxury for the rich. These early markets would encourage the entry of new service providers, and support cost reductions from economies of scale. Eventually, millions might share the experience. Air travel has expanded to providing a service for millions of people in less than 40 years. Today, the system is a US $ 250 billion industry, with roughly a million people in the air at any given time. But there are only a few astronauts in space at any one time.

4. The International Space Station Program

What effect does a government funded Space Station have on the future in space? The government is investing large amounts of money into Space Station technologies, components, life support systems, etc.. These directly apply toward future space hotels and facilities. But the non-commercial way in which these technologies are developed strongly affects the transfer of these technologies to private projects. Either the company does not have full rights to the intellectual property, or the information is disclosed to competitors. Technical solutions are selected with a consideration of political factors rather than of cost or performance.

Building a Space Station for political reasons, and then presenting it to the public as solely a research facility, causes the government to slide into circular arguments. The government presents the Space Station as the answer to research needs, and then sponsors research – as long as that requires the Space Station. The lucky researchers whose interests match the government's receive government support and access to the costly facilities. They do not see this money, they do not control it, and no alternative Space Station or other facility can be rewarded with their business. Scientific users cannot act as Space Station customers. ISS systems are built to satisfy the true customers of ISS – the politicians – who are often more concerned about *where* the work is performed than about performance or cost.

For the developers and operators of the government's Space Station, what is the natural reaction toward the vision of public access to space? Yes, this will be possible – some day. We all say that we want to see low cost access to space – but do we? How would public officials react if a second Space Station were built at a fraction of the cost of the ISS? Many people have devoted their careers to explaining why the ISS is so expensive – it is so difficult that it can only be done by the combined efforts of many governments. If a low cost option appeared tomorrow, who among the current space industry would benefit? Could ISS suppliers sell their next set of modules at a small fraction of today's cost without upsetting their billion-dollar government customer? Both government and industry are strongly motivated to maintain the *status quo*. When the ISS is described as a step toward a future of public space travel, it is with the built-in assumption that this dream will fail.

5. Changes to the ISS Philosophy

Space tourism does not require the ISS program. But the ISS has an opportunity to encourage the development of the future of large-scale public access to space. The objective of the current ISS should remain as a research facility, but the way in which the facility is operated will have a strong effect on the future of humans in space. The central change required in the program is a change in perspective. The ISS must be operated with the assumption that there will be a future for humanity in space. The long range perspective should not be a 'dozen astronauts in space', but hundreds, thousands, and eventually millions of people in space. After a hard struggle for the current ISS, it is difficult to imagine this kind of future. But it is essential to guide human space activities today.

5.1 Scientific Research

The biggest concern for any changes to the ISS program is the risk of disturbing the primary research missions of the facility. This paper does not propose distorting the process of selecting scientific experiments on the ISS. The proposal is to incorporate a vision of future public access to space as part of the context of setting the research priorities.

The current view is that ISS human research is needed to prepare for sending humans to Mars, which provides a context for the evaluation of proposals. In the same sense, the acceptance of a future with public access to space would not re-orient ISS research, but could provide a context to encourage related research. Many objectives overlap those for sending humans to Mars, including a more complete understanding of the effects of microgravity, animal studies at intermediate g levels in centrifuges, radiation effects, long term habitability factors, noise levels, etc.. Continued development of medical care in space also relates to both visions of the future. A basic factor in human research on the ISS should be the assumption there *will* be a future with more people in space.

Preparation for public space access must also include the study of space effects on a broader cross-section of humanity. John Glenn's space flight at age 77 illustrates an expansion of the age distribution. Public access to space will test many other limits beyond the composition of the current astronaut corps. Medical constraints on astronauts are based on the need to perform mission-critical activities. Is there some way that research on the ISS can extend the database of health effects, without increasing risks to others, or reducing ISS effectiveness? Perhaps there will be opportunities for guests on re-supply flights, or other mechanisms for flying a wider range of individuals. Perhaps there are animal analog experiments to address these questions. Without the ISS, these questions could be addressed on future private facilities. The ISS simply offers an opportunity to advance the schedule.

5.2 Technology Development

Future space hotels will benefit from the development of many Space Station technologies. Future facilities will need to have a similar mix of subsystems and hardware. Hardware and tools developed for life support and to enable the crew to work will be important for new facilities. Many of the ISS design goals match the needs of future facilities – routine operation, high reliability, on-orbit maintainability, low levels of maintenance crew time, and simple interfaces. Future space facilities need to apply the lessons learned from the current design, and increase the emphasis on low operational costs.

The current program will continue to require new equipment for replacements and additional capability. To encourage space industry development toward a larger future, the new equipment should be purchased commercially to the greatest extent possible. The government should identify its needs and (simplified) space qualification requirements, but not specify the design solution. Companies should be able to sell their equipment and retain ownership (and proprietary information) about their designs.

5.3 Policy and Programmatic Issues

The ISS program has struggled for a clear identity and role. The needs of many users were considered in the compromise design. Among many potential users, no one wanted to be labeled as the primary reason for the ISS. Eventually the main role was identified as a microgravity research facility, and as the next step in a continuing, but undefined, plan for government exploration of the solar system. Meanwhile, the public wonders when they will be able to participate in space. Impatience in the U.S. Congress led to policy statements that ISS should have, as its objective, the stimulation of commercial activity in space. NASA policy now includes similar language [Reference 11]. What are the policy and programmatic implications of these new objectives? The ISS can support commercial space activity by providing facilities for commercial users. But the ISS can also operate and evolve the facility using commercial suppliers.

The encouragement of commercial space activities aboard the ISS necessarily involves some release of governmental control. If an ISS support service is to be acquired in a commercial fashion, the government will not have the access and oversight to which they are accustomed. The government will need to return to a role of the customer which it plays in many other areas. It communicates needs and standards, and selects among commercial solutions. Can the ISS operate like this? In some sense, this program has been learning to live in this kind of environment. The ISS has many international partners, which has forced NASA to operate in an environment where they do not have complete control. The introduction of purely commercial suppliers would not be that much of a change. The program needs to learn how to define standards and accept solutions based on those standards. Businesses cannot invest in innovative solutions, if there is a risk that the government can change the rules to pick a winner.

5.4 Space Economics

The current ISS program, and future space facilities, will benefit from the introduction of commercial practices into space operations. The actual costs of conducting current space operations must be made more visible if we are to develop improved, lower cost systems. Resource allocation between government supported users on the ISS could follow other examples (e.g., Jet Propulsion Laboratory spacecraft), and apply an 'internal market' to barter excess resources. As ISS operations shift toward commercial operations, price labels could be applied to this barter system, providing a competitive standard for ISS service providers to develop lower cost options. The ISS could stimulate the creation of new industries – if the government gets out of operations, purchases commercial services, and allows private ownership of assets on the ISS.

6. Summary

The development of space tourism does not depend on the ISS. Space tourism is proceeding with its own momentum toward a future far larger than a single research facility. But the larger future for humanity in space could be enhanced by changes in the ISS program. The research and technical changes are minimal; there is no need for a significant impact on the primary mission of the ISS. The ISS must simply be operated and evolve with the assumption that there will be a future for people in space. It must operate in a new model of purchasing Space Station support services and capabilities from private commercial suppliers. The government facility can then be the stimulus for an emerging industry, instead of a force preserving the *status quo*. These changes offer near term enhancements for the ISS facility, as well as long term benefits.

References
1. Space Publications: *State of the Space Industry 1999*, www.spacebusiness.com, Bethesda, Maryland, May 1999
2. Space Adventures: *1999/2000 Program Catalog*, www.spaceadventures.com, Arlington, Virginia, March 1999
3. Wright, J.W. (editor): *The New York Times 1999 Almanac*. Penguin, New York, 1998
4. *Space Tourism – The Story So Far*, www.spacefuture.com/tourism/timeline.shtml, May 1999
5. Whitehouse, D.: Hilton to back space hotel, *BBC News Online*, news.bbc.co.uk/hi/english/sci/tech/newsid_293000/293366.stm, London, March 9, 1999
6. Nuttall, N.: Space odyssey becomes reality, *The Sunday Times*, www.sunday-times.co.uk/ news/pages/tim/99/05/01/, London, May 1, 1999
7. Berger, B.: U.S. Developer Sets Sights on Space Tourism, *Space News*, Vol. 10, No. 20, www.spacenews.com, Springfield, Virginia, May 24, 1999
8. Clarke, A.C.: *2001, A Space Odyssey*, Signet Books, New York, 1968
9. World Travel Tourism Council, cited in: Coniglio, S. M., *Practical Tourism in Space*, www.magicnet.net/~sam123/spacetou.html, 1996
10. Collins, P., Stockmans, R., and Maita, M.: *Demand for Space Tourism in America and Japan, and its Implications for Future Space Activities (1995)*, www.spacetourism.com/ tourism/timeline.shtml, May 1999

11.   National Aeronautics and Space Administration: *Commercial Development Plan for the International Space Station*, <www.hq.nasa.gov/office/codez/policy.html>. Washington, DC, November 1998

International Space Station Commercialization: Can Canada Blaze the Trail Forward?[1]

A. Eddy, A. Poirier, Canadian Space Agency, 6767 Route de l'Aeroport, St. Hubert, Quebec J3Y 8Y9, Canada

e-mail: andrew.eddy@space.gc.ca,   alain.poirier@space.gc.ca

Abstract
All of the partners involved in the utilization planning for the International Space Station (ISS) have announced their desire, in some form or another, to "commercialize" a portion of their ISS utilization rights. Although commercialization may mean many different things, the ISS partners are confronted with common problems that block the path towards commercialization. This paper addresses what Canada plans to do to move forward with its commercial program.

1. Introduction

The first point to be addressed by anyone meaning to discuss commercialization is what commercialization is. In Canada, the ISS Commercialization (CZ) program does not aim to transfer technologies developed for space to other sectors or to facilitate access for new user communities to the ISS. The ISS Commercialization Office has a mandate to identify new partners to co-finance the utilization of the ISS. These partners are from both private and public sectors.

It is widely recognized that the ISS utilization "market" is at best embryonic and volatile. While completing the assembly of the ISS will undoubtedly improve this situation, most in the utlization community agree that other hurdles make ISS commercialization difficult to envisage in any near future.

The most common hurdle cited is the exorbitant cost of access to space. While we would readily agree that a substantial reduction to the cost of access to space would serve as a substantial impetus for commercial activities on the ISS, in our view, the principal barriers to the commercialization of ISS are of a different nature. They are awareness, culture and partnership philosophy. While little can be done to address the high cost of access to space, it is possible to act in relation to the three barriers identified above.

---

[1] This paper is the sole responsibility of its authors and does not necessarily reflect the views of the Canadian Space Agency or the ISS partner nations and agencies.

G. Haskell and M. Rycroft (eds.), International Space Station, 79-82.
© 2000 *Kluwer Academic Publishers.*

2. Barriers to ISS Commercialization

2.1 Awareness

One of the most surprising facts that one discovers in discussing ISS Commercialization with potential user communities is an astonishing lack of awareness with regard to what the ISS, and microgravity research in general, have accomplished and may offer in future.   For over ten years, the international spacefaring community has touted the commercial potential of research undertaken on the US Space Shuttle, on the Mir space station, and on free flying platforms. Around the world, many nations have drop towers and fly parabolic flights on dedicated aircraft to prepare for research in space. Suborbital rockets offer minutes of research ; returnable capsules offer weeks or sometimes months. Yet despite this impressive infrastructure and the resources dedicated to commercial success, space managers are hardpressed to identify any specific breakthrough attributed directly to space research.

2.2 Culture

The problem of awareness in fact hides another, more serious challenge — a culture gap between the private and public sectors. Whatever the intrinsic value of space research, because of the governmental infrastructure involved and the implication of agencies from around the world, space research remains focussed on long-term applications. Its very structure cannot adapt to short product development cycles and the need for regular, timely results. Space research remains focussed on public sector and academic needs ; it does not adequately address the needs of the private sector.

2.3 Partnership Philosophy

Finally, diverging partnership philosophy is a real hurdle to success for ISS CZ. Every solid partnership is characterized by its win-win outcome. Different partners bring different strengths to the partnership, and take from it different things. Typically, the public sector has broad goals affecting the public good, and specific scientific and technological objectives. The private sector has a simple, specific goal, exacted by shareholders — profit. Both partners must recognize their differences, and accept each others' different needs.

3. Overcoming the Barriers

3.1 Awareness: Spreading the Good News

Awareness is in many ways the easiest barrier to overcome.   The excitement generated during the launch of the ISS's first elements and during the subsequent assembly sequence has already improved knowledge of what the ISS is and will continue to do so. Space agencies around the world have begun to organize awareness and education activities that will build on this excitement. With the first research on the ISS scheduled to begin by late 2000, much of the initial awareness work will be accomplished.

It is crucial, however, to ensure that the work does not stop there. First, space agencies must ensure that the news which they bring to the world is good news, not news of delays, of difficult research conditions or of science in search of applications. These points we will address shortly.

Secondly, it is essential that the detailed message of the specific relevance of space research to specific communities be brought to them in a targeted fashion.   This entails leaving the traditional spacefaring community and speaking to new users. This is something that space agencies have traditionally found difficult to do, and to which they must dedicate particular attention.

3.2 Culture: Doing Business Differently

The differences between government and the private sector, and the ways in which they conduct their business, are perhaps the most important barriers to ISS commercialization. In fact, to refer to the business of government is an oxymoron. Government is not actively conducting business, at least in the sense meant by industry. Consequently, government cannot understand quarterly outlooks, or the important impact which flight delays and program overruns can have on stock value or financing arrangements.

If space is ever really to become business, government must accept that it has to step out of the way and let business take the lead. In the case of ISS commercialization, however, until greater success is demonstrated, industry is unwilling to assume all the risks. Yet if government assumes the risks, and industry the benefits, this is essentially a government subsidy of shareholder profit. It is of course unacceptable to government and to the public at large.

In order to overcome this hurdle, government must create the conditions of a market in which free enterprise is able to take hold.   For space commercialization, this means regular, timely, affordable access to space.

Regular access is currently planned for the ISS. The three month increment, while far from perfect, probably offers sufficient regularity to meet early industrial needs. Furthermore, several private sector companies are considering more regular access both to and from the ISS.

Timely access involves rethinking what must be done before a payload goes into space. This will be one of Canada's priorities in the coming months. In order to meet commercial needs, we believe that the payload manifest cycle should not exceed 12 months.

Finally, affordable access can be possible, even within current constraints. While little can be done to reduce the cost per kilogram to orbit, much can be done to decrease the size and weight of payloads to make the absolute cost of private investment reasonable to serious entrepreneurial consortia.

4. Conclusion

For the limited purposes of this paper, it is useful to conclude with what Canada aims to do to bring ISS Commercialization closer to reality.

The Canadian Space Agency has decided notionally to allot 50% of its utilization rights to the commercial program. Different approaches have been adopted for internal and external space, given their different natures. The Canadian Space Agency has also decided to continue pursuing non-traditional commercial applications such as entertainment and advertizing. These programs will be open to international industry, in partnership with Canadian companies.

If the Canadian Space Agency is successful in creating an environment conducive to commercial activity within its limited allocation, other partners may choose to make portions of their allocations available under similar conditions, ultimately leading to fully-fledged commercial activity on the ISS.

Transitioning to Commercial Exploitation of Space

D. Hamill, M. Kearney, SPACEHAB, Inc., 1331 Gemini Avenue, Houston, TX 77058, USA

e-mail: dhamill@spacehab.com, kearney@spacehab.com

Abstract
In the 1960s and '70s, humanity entered space to explore its novel environment. In the '80s and '90s, human presence in space focused on experiments which probe larger scientific boundaries. The decades ahead should see manned space exploited for economic benefit. But the structures that served the space program well for exploration and experimentation may not meet the needs of commercial exploitation. Commercial users of ISS will demand low cost, short timelines, and assured access for the services which they purchase. Space activities must be integrated into the normal flow of international commerce. This paper discusses what potential commercial customers expect from their use of space. It highlights the differences between what exists and what is needed, and outlines an approach to transitioning to a system which can mesh with the established mechanism of industrial capitalism. Commercial manned space will be considered successful only when sending work to laboratory based in space is as unremarkable as sending it to a laboratory in another city.

1. Introduction

Although the space program still strikes most of us as esoteric and novel, people have been in space now for almost four decades. Nor is space the first new environment to be opened to humanity in the Twentieth Century. Without particularly planning it, the opening of space is following roughly the same pattern of development as its two major predecessors, the atmosphere and the subsea. We can therefore lean on these models to move with confidence into the next logical step in the opening of space — its exploitation for economic benefit.

This paper argues that the transition to commercial exploitation can be accomplished most easily by allowing established commercial mechanisms to come into play. Space is not so inherently different that it demands inventing new structures for commercial operations. Market forces promote the best practices and weed out the worst ideas by their very nature. Once the environment for commerce has been established, efficient mechanisms will drive the details of commercial space operations down paths that no one can foresee at this time.

2. Looking Back to See Ahead

When humanity undertakes to conquer a new environment, whether it is the atmosphere, the sea bottoms, or outer space, the first decade or two are characterized as *exploration*. During the exploration phase, both the

G. Haskell and M. Rycroft (eds.), International Space Station, 83-90.

environment itself and the technology that open that environment are largely unknown and therefore dangerous. Operational procedures are improvised, and revised as problems occur. Early aviators died from structural failures and aerodynamic effects that were not anticipated; early divers were killed or injured by depth effects and equipment failures. The space program, though no less bold, had fewer casualties in part because it applied lessons of caution from the past. Nonetheless, dangerous near-misses and even tragedy marked the first decades of manned space flight as clearly as the exploration of the air and subsea.

By the end of the exploration phase, design fundamentals are established and critical hazards conquered. Procedures, though still immature, support routine operation; a rudimentary infrastructure is emerging. The next two decades or so may be characterized as the *experimentation* phase, during which the capabilities and limitations of the new environment are probed. Technical advances, a growing understanding of the environment, and evolving operational procedures reduce the personal danger to a point that is tolerable by most people. Experimentation identifies and evolves, though it does not optimize, the hardware that supports the best applications of this environment. The infrastructure, also not optimized, is robust and capable.

The space program today stands at the end of the experimentation phase. If history and the stated intention of our space leaders prove true, the next phase of opening the space environment will be *exploitation* of space for its economic benefits. This phase is already underway for non-manned space. The exploitation phase is characterized by refining and optimizing those opportunities that have been identified and developing new activities enabled by these efficiencies. The reliability and cost reductions demanded by customers evolve new hardware and modes of operating. The next twenty years, and on to the indefinite future, should see space routinely used for economic purposes until it becomes as integrated into the world economic structure as the airlines and off-shore oil extraction.

3. Economic Activity in Space

3.1 Early Opportunities

The experimentation phase of the space program has uncovered potential commercial uses for manned space. In general, these have a very high value per unit mass. Protein crystal growth is the archetypal example. Proteins crystallized in microgravity have a high value to the pharmaceutical industry if they cannot be satisfactorily grown on Earth yet provide key insights on the road to developing valuable drugs. The raw materials for protein crystal growth are extremely light, and the mass of the support equipment, the growth cells and refrigerated incubator, is modest and can be shared by a large number of customers. Protein crystal growth is the only application that has proven its cost-effectiveness to commercial users in the cost environment of the American Space Shuttle.

The International Space Station will improve the cost-effectiveness of applications with a low sample mass but a high equipment mass. Furnace applications, for example, have modest sample and container masses but high equipment masses. Significant commercial markets for inorganic crystal growth and data on the thermophysical properties of metals will emerge only when the launch mass of the furnace can be shared by more users than can be accommodated by a single Shuttle sortie. Once the equipment is on orbit, only the samples will have to be lifted.

Behind these applications stand several more whose commercial viability waits for reductions in either the equipment mass to orbit or the cost of mass to orbit. Mass reduction will require the discovery of a market potential that will propel interested capital to invest in new equipment generations. Reductions in the cost to orbit will require patient evolution of new approaches to space operations. Transgenic manipulation of plants serves as an example of a potentially lucrative commercial area that still requires significant recurring launch mass.

3.2 Generic customer requirements

Although cost reduction is the *sine qua non* of commercial viability, other factors must also change on the road to space exploitation. Commercial customers will demand faster service and more schedule certainty from their suppliers than the space program is currently capable of providing. The authors have held extensive discussions with the commercial user community about their needs. Table 1 summarizes the difference between what exists today in the STS-ISS infrastructure and what commercial users will demand from the providers of space services.

Besides these quantifiable improvements, commercial users will insist upon absolute protection for proprietary information. Furthermore, they will not tolerate documentation overhead greater than they develop for comparable testing at a terrestrial laboratory. Commercial users will not go to space to say that their product came from space but because it makes cold business sense to do so: space *per se* adds no value to the product. In short, commercial users of space wish to treat activity in space as though it were at any other laboratory with some unique capability.

	Where we are today	Where we need to be
Cost	• $20,000 / kg to orbit • Certification paperwork adds 5x to cost [Reference 1]	• < $10,000 / kg to orbit • Nominal additional cost to operate in space
Timelines	• 2-3 years from commitment to flight • > 24 month flight integration template	• 0.5 year from commitment to flight • ~ 1 month turn-around
Schedule uncertainty	• ± 3 months at date of commitment	• ± 1 week at date of commitment

Table 1. Customer demands for commercialization

4. The Coming Paradigm Inversion

We should expect that the structures – hardware and operations – that will grow out of the exploitation phase will be as different from those developed in the experimentation phase as its structures were from its predecessor. Each phase responds to its own imperatives, so the structures that served the experimentation phase must be modified to serve exploitation's imperatives: cost sensitivity, timeliness, assured access, process simplicity, and protection of proprietary data. At this juncture, we must resist the temptation to try to predict what those structures will be or should be. The structures to support space exploitation must evolve over time as their predecessors did. Transitioning to commercial exploitation, then, is not a question of putting new structure in place but *establishing the environment in which the structures can evolve.* An environment that fosters commercial activity will inherently take advantage of existing commercial structures when possible, and invent new ones only when required.

In the right environment, commercial markets grow by their own dynamic (Figure 1). The price point for products and services will attract a certain customer interest. Vendors will seek ways to better serve their customers and reduce costs. These improvements permit them to expand their markets. This resets the price point and moves the process into the next cycle

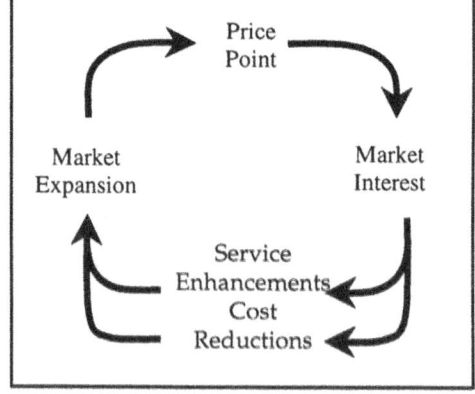

Figure 1. The Capitalist Dynamic

In the current structure, the price point is set by the costs, namely the supplier costs, profitability, and the costs of infrastructure. The price point then dictates the value of the scientific work done. Despite the very high cost of space operations, this structure has worked because the customer, the governments, place intangible value on developing space, and that value supplements the tangible value of the research to make it worth the cost.

In the commercial environment, the value is established by the competitive market. *Value*, not cost, establishes the price point. The vendor must trade his supplier and infrastructure costs against profitability at that price point when deciding whether and how to serve the market. The *vendor*, rather than the customer, drives this paradigm.

5. Priming the Capitalist Dynamic

Because profitability, supplier costs, and infrastructure costs are key to priming the capitalist dynamic, we examine each separately and recommend how to establish a conducive environment for it.

5.1 The Profit Motive

The century-long competition between capitalism and communism showed that, whatever its shortcomings, capitalism is unequalled for creating wealth. The desire for personal aggrandizement provides the capital that develops new products and services to the benefit of the whole economy. In a competitive environment, profit motivates efficiency and innovation in a way that good intentions alone cannot match. It also propels companies to improve and expand existing business lines, identify and develop untapped markets, grow demand for its products, invest in areas that have good prospects, and withdraw support from unproductive areas.

The profit motive functions poorly or not at all on a fixed-fee, cost-reimbursed contract. During the experimentation phase, while the government bore the risks, a profit capped at a percentage of cost did not discourage suppliers. However, in a normal commercial environment, the vendor must bear the financial risk and therefore cannot be artificially limited in profit.

To create the proper environment for commercial exploitation, the authors suggest that the following maxim be adopted throughout the space program: "anything that can be done commercially should be". Certain Space Station functions, like the selection and prioritization of non-commercial research, must necessarily be done by disinterested parties. Most other functions could be commercialized. Under this maxim, approval for commercial proposals would be presumptive, depending only on meeting three tests:

- the customer receives equivalent or higher value for the product or service,
- the fixed price for that value is equivalent to or lower than the non-commercial baseline, and
- the product or service is based on a significant amount of private capital invested at risk.

To implement such a major change, commercialization must have powerful advocates inside the space agencies. Commercial proposals often cut across the interests of different departments within an agency and occasionally put those interests in conflict. The advocate must be in a position to reconcile such conflicts. Although most commercialization proposals should come from private initiative, the advocate should also proactively identify functions that are ripe for commercialization, but these should be based on technical maturity and profit potential, not funding shortfall. The sounding rocket programs, for example, would be an excellent target for commercialization.

5.2 The Cost Structure

Today's cost structure and operational procedures for space result from 40 years of public management. Spending the public's money brings special considerations, such as procurement regulation, national boundaries, and politically mandated set-asides, that do not apply to private spending. Moreover, the government insures itself by a significant documentation burden that multiplies the costs of hardware five-fold [Reference 1]. The cost-based pricing necessary during exploration and experimentation does not inherently control costs.

Making a profit in a competitive commercial environment is inseparable from improving efficiency. Cost reductions improve profitability, which generates capital to reinvest in new products and services, and permit price reductions that increase demand. Shortened timelines open the market to more customers. The first step, then, in reducing the cost and timelines of work in space is to implement the maxim above: "anything that can be done commercially should be".

For example, commercial practice includes mechanisms and models that could reduce the documentation burden that adds so much cost and time to space flight. In a commercial environment, insurance and partnership with insurers reduce hazards and underwrite any damage that results from negligence. Commercial organizations like Underwriters Laboratory make a customer-friendly business of certifying that hardware complies with established standards. Workman's compensation, backed up by tort law, enforces safety without the need for massive documentation. Other ideas would surely arise in an environment that encourages commercial take-over of the routine activities of space.

5.3 Infrastructure

Infrastructure supplies the foundation for commercial activity. The Space Transportation System, with its associated launch and mission control facilities and communications network, along with the International Space Station and its resources, constitute the existing infrastructure for manned space. Soon the HTV, ATV, and their launch and support facilities will join it. Establishing and maintaining infrastructure is traditionally a government responsibility. It can be very expensive and, though it contributes broadly to economic well being, may not have a tangible return. Generally the government does not try to recoup the cost of building the infrastructure but may charge a "user fee", such as a toll or a gas tax for highways, to support operation and maintenance. Such a fee offsets but does not ordinarily recover all operating costs. Because these fees are an expense that affects a vendor's profitability, they may be set low or waived entirely to encourage a struggling new industry.

The value of the infrastructure cannot be assessed while its supply exceeds the commercial demand for it, as will probably be the case for the first several years of ISS operation. Assigning a value based on cost – the old paradigm – will distort the whole valuation process, especially while the costs are a vestige of the experimentation phase. Only commercial competition will establish a market value for infrastructure. Vendors must have time to build market demand to a point that equals or exceeds supply before competition for scarce resources can establish an appropriate value for infrastructure. This valuation, then, can serve as the starting point for vendors to determine whether and how to provide commercial infrastructure.

The government will have to bear the responsibility for infrastructure until the market valuation approaches the current costs, though it may expect to offset some of its marginal costs. Once the profit potential of infrastructure becomes realistic, the "whatever can be done commercially should be" maxim will mandate its commercialization. If history is a guide, the transition to commercial operation might begin as a regulated utility with a single supplier permitted to make a reasonable commercial profit under oversight that prevents price gouging. Once the market demand is large enough that other competitors can capitalize a competing infrastructure, as has happened in the last decade in telecommunications, the government may deregulate and leave the business entirely.

History gives us every reason to believe that this evolution will happen spontaneously. Trying to hurry it artificially in order to recover costs for other space exploration can only stifle commercialization by adding costs above those that the market can bear. The way to accelerate the handing over of infrastructure responsibility is to encourage market building by commercialization as much as possible as soon as possible.

6. Handing-over between Paradigms

The old and the new paradigms can coexist as long as they are not forced to interact. The commercial allocation of ISS must be allowed to evolve value-sensitive, vendor-driven structures outside the structures and contractual arrangements that support cost-based, customer-driven scientific utilization. However, we foresee a day when the scientific community may wish to take advantage of the efficiencies developed by the commercial side. This should be done in a way that encourages commercialization. For example, once an infrastructure value has been established, research grants could include funding to cover that cost, funding shifted from the operations to the utilization budget. This would allow researchers to choose between the non-commercial and commercial access to space, providing a larger market for the commercial vendors while giving researchers use of the facilities and processes available on the commercial side.

Ultimately, the governments' role in microgravity research may be reduced to providing grant money that researchers budget for commercial services in space in the same way that recipients of other government grants budget for terrestrial laboratory services.

7. Conclusion

Once it is agreed that "everything that can be commercialized should be", obstacles *will* fall, markets *will* develop, and economic benefit *will* begin to flow from space. The only genuine barrier to manned space commercialization is failure – even by its proponents – to choose commercial alternatives over familiar ones.

References
1.   *Reducing the Cost of Space Infrastructure and Operations*, NISTR 5255.

International Space Station Commercialization Study

J. J. Richardson, Potomac Institute for Policy Studies, 1600 Wilson Blvd., Suite 1200, Arlington, VA 22209, USA.

e-mail: jrichardson@potomacinstitute.org

Abstract

In 1996 the Potomac Institute for Policy Studies performed a study on commercializing the International Space Station (ISS). The work was principally funded by the National Aeronautics and Space Administration (NASA). The Institute collected and analyzed publications and sought extensive counsel across industry and government. Beginning with our panel, chaired by Mr. James Beggs, a former Administrator of NASA, we interviewed over 200 people, representing approximately 50 companies, universities, and government agencies. We also conducted 12 case studies to look at the potential utilization of piloted space flights in Earth orbit.

The study suggests that commercialization of human orbital space flights could yield considerable benefits. Although there are some plausible commercial space-based ventures, we found no corporations that could access space without government help. The amount of help needed from NASA is considerable; we found that successful ISS commercialization demanded a broader context than the station itself, involving space access and other orbital resources. In the face of this, we found that NASA had articulated considerable support for commercialization, but had failed to commit the attention and resources needed to make it happen.

1. Purpose of Study

The International Space Station Commercialization (ISSC) Study was performed by the Potomac Institute for Policy Studies (the Institute), principally under a grant from NASA [Reference 1].[1] The Institute and other companies also provided financial support. Views expressed are those of the Institute and are not necessarily endorsed by NASA or the other contributors.

The objectives of the study were to present independent, informed and updated perspectives on three questions pertaining to the commercialization of human orbital space flight, and in particular the ISS [Reference 2]. Its findings rest upon the assumption that NASA will deploy the ISS within the next six years. The questions asked were as follows:

- Are there compelling potential benefits from the commercialization of human orbital space flight?
- Are there viable areas of opportunity and plausible commercial ventures?
- What, if any, should be the government's role in fostering commercialization?

---

[1] This report is also accessible on the Institute's website, www.potomacinstitute.org.

G. Haskell and M. Rycroft (eds.), International Space Station, 91-96.
© 2000 *Kluwer Academic Publishers.*

During the course of the study we contacted over 200 people, representing approximately 50 companies, universities, and government agencies. We convened the Space Commercialization Experts Panel, the members of which are given in Table 1:

Member	Selected Experience
Mr. James Beggs, Chairman	President, MAKAT, Inc. Former NASA Administrator Former Executive Vice President, General Dynamics Former Deputy Secretary of Transportation
Dr. John McLucas	Former Chairman, NASA Advisory Council Former President, COMSAT General Former Secretary of Air Force Former Administrator, FAA
Mr. James Rose	Former Assistant Administrator for NASA's Commercial Programs
Mr. Howard Schue	Partner, Technology Strategies and Alliances Corporation
Dr. Terry Straeter	President and Chief Executive Officer, GDE Systems, Inc.

Table 1. Members of Space Commercialization Experts Panel

We also conducted the twelve case studies listed below.

- Case Study 1. Space Hardware Optimization Technology (SHOT)
- Case Study 2. Boeing: Mir Experience
- Case Study 3. Microencapsulation (Vanderbilt University)
- Case Study 4. Macromolecular Crystallography (University of Alabama in Birmingham)
- Case Study 5. NASA Space Sciences Laboratory (Marshall Space Flight Center)
- Case Study 6. Centers for Casting and Power and Advanced Electronics (Auburn University)
- Case Study 7. New York City Economic Development Corporation
- Case Study 8. Zeolites (Worcester Polytechnic Institute)
- Case Study 9. Virtual Presence (LunaCorp)
- Case Study 10. Gallium Arsenide (Space Vacuum Epitaxy Center and Space Industries, Inc.)
- Case Study 11. X-Ray Device (University of Alabama in Birmingham)
- Case Study 12. Education Programming (Walt Disney Imagineering

2. Summary of Conclusions

The results of the study convinced us that commercialization of human orbital space flight could offer significant benefits to NASA and the nation.

Benefits to NASA's mission include:

- Better and more affordable space assets
- Increased utilization of the Space Shuttle, ISS and Reusable Launch Vehicles
- Release of NASA resources for application to new science frontiers
- Leveraged private investment
- Improved innovation and importation of commercial technology to space endeavors
- Increased public support for space operations.

Three national benefits identified were:

- Enhancement of U.S. industry competitiveness
- Spin-offs of new technologies to non-space industries
- National prestige.

We also found interesting and plausible space-based commercial ventures.

- The most viable opportunities lie in the privatization of government functions, such as resupply and operation of the ISS
- Emerging privatization opportunities encourage industries to develop better and more affordable operations, services, support, and space equipment. Importantly, this also enables industry to better serve commercial space ventures
- Commercial research ventures, in biomedicine and materials, provide important insights into Earth-based processes
- Near-term commercial opportunities exist in education, entertainment, and advertisement.

However, no commercial venture was able to get into space without help from the government. Major problems cited included high launch and operation costs, low flight frequency and reliability, long launch lead times, and expensive indemnification against flight failure. Government help in situations like this is consistent with historical precedents set during the initiation of U.S. transportation systems, such as canals, rail, air, and interstate highways.

NASA had indicated a desire to transfer ISS and other human orbital space flight activities to the private sector [References 3 and 4]. They had also agreed with the concept of offsetting NASA's expenses through a healthy commercial market. Even so, NASA's efforts to foster commercialization were declining. NASA's superb accomplishments in space science continued, despite diminishing manning levels and budgets. But, in the inevitable tradeoffs between mission areas, commercialization seemed to be losing. For example:

- The percentage of NASA's budget dedicated to commercialization has declined steadily since 1993. At its highest, this portion was still less than one percent
- Reorganizations left NASA without an institutional center to accommodate commercial participants
- NASA lacked a coherent outreach program to commercial business
- Many publicly-stated promises went unfulfilled
- Although procurement and procedural inflexibilities have been reduced, they are still too typical of NASA's operation.

Under these circumstances, the corporations contacted tended to assume that space access would remain too risky and subject to bureaucratic processes. This has stifled creative thought about space utilization in corporate boardrooms around the country, and posed a serious detriment to commercialization.

3. Recommendations

We suggested a strategy of privatization-to-commercialization of human orbital space as a logical means of achieving NASA's aims. Such a strategy will not be an easy undertaking. It will demand enthusiastic follow-through, with active support from the highest echelons of NASA. Many components of our recommended strategy are reflected in NASA's recent ISS commercialization plans.

There must also be an implementation arm to create a more innovative and productive link between NASA and the private sector, and to develop and husband supporting policies, directives, and strategies. Some characteristics of the proposed strategy are:

- Clearly stated commercialization goals, with a focal point within NASA to pursue them effectively
- Private sector representation in formulating plans, strategies, and policies, which should include an outreach program to convince commercial industry of the viability of operating in space, from both a technological and business perspective
- Compelling incentives for NASA management and personnel to support and accomplish commercialization goals
- A "Privatization-to-Commercialization" approach, with sufficient NASA investment to support it. This approach must mandate the use of

privately developed infrastructure by outsourcing and discouraging in-house competition with the private sector. It must support the use of privatized facilities for commercial ventures and a realistic return on equity for the private sector, considering risks. Where appropriate, NASA should accept the role of anchor tenant

- A policy of providing support, encouragement, advice, and space access to diverse commercial sectors
- Added emphasis on reducing impediments to more frequent and affordable space access.

It was suggested that a Commercial Development Office (CDO) and a Space Economic Development Corporation (SEDC) should be established by NASA. Some key points are as follows:

- The need for commercial advocacy within NASA is sufficiently compelling to warrant changes in organizational structure. First, the CDO should serve as a focal point and advocate commercialization within NASA. The CDO should then organize a public/private partnership SEDC, which would take over some of the functions of commercialization and, eventually, most of the commercialization effort
- The CDO would begin this process by refining NASA's strategy, developing contacts within the private sector, consulting with NASA Offices and Field Centers, recommending some early policies, and developing innovative approaches to privatization. The CDO should contain sufficient governmental expertise to coordinate actions and obtain support from within NASA. The major thrust of the CDO, however, would be business; therefore, it must include personnel with extensive experience in the business world. Venture capitalism, business and legal processes, as well as technology and product development must be represented. The staffing for the business side of the CDO should be found from outside the government. Such people would also help to form the SEDC
- The SEDC would represent the link with the private sector, providing a business environment to those industries seeking access to space for commercial purposes, or to those interested in privatization of space assets. It would begin as a quasi-government corporation. Its mission should include forming consortia, negotiating business agreements, formulating venture plans and strategies, and performing other functions that government cannot accomplish. The SEDC could accept funds from government or the aerospace industry. Large space assets ventures, such as the Reusable Launch Vehicle (RLV), could form their own development corporation, or rely on the SEDC. This organization would eventually lead the commercialization effort, acting in the role of a true development

corporation. Until this "spin-off" occurs, they would support the CDO in conducting a series of outreach programs, encouraging industry to consider human orbital space flight, reaching a better understanding of the special problems of the private sector, and exploring the benefits of space to the commercial marketplace. The SEDC would also help NASA to become more appreciative of private sector values and approaches.

References
1. Potomac Institute for Policy Studies, *The International Space Station Commercialization Study* (PIPS-97-1), March 1997
2. Richardson, J.: *Study Plan for the International Space Station Commercialization Study*, PIP 96-5, 8 August 1996
3. General Accounting Office, *NASA Infrastructure: Challenges to Achieving Reductions and Efficiencies*, September 1996
4. Congressional Budget Office Memorandum, *Budgetary Treatment of NASA's Advance Commitments to Purchase Launch Services*, June 1995

Selected Bibliography
1. NASA Commercial Programs Advisory Committee, *Charting the Course: U.S. Space Enterprise and Space Industrial Competitiveness*, 1989. NASA Advisory Council Task Force on International Relations in Space, International Space Policy for the 1990s and Beyond, 12 October 1987
2. National Research Council, *Engineering Research and Technology Development on the Space Station*, 1996
3. Boeing, Martin Marietta, General Dynamics, McDonnell Douglas, Lockheed, and Rockwell, *Commercial Space Transport Study Final Report*, April 1994
4. Report of Aerospace Research and Development Policy Committee, Institute of Electrical and Electronics Engineers, *What the United States Must do to Realize the Economic Promise of Space: Who Would Build a Second Space Station?*, 1993
5. Space Studies Board, National Research Council, *Microgravity Research Opportunities for the 1990s*, 1995
6. Handberg, R., *The Future of the Space Industry*, Quorum Books, 1995
7. Boeing / Peat Marwick Commercial Space Group Report to NASA, *Services to Support the Commercial Use of Space*, 1988.
8. Harr, M., et al.,: *Commercial Utilization of Space*, Battelle Press, 1990
9. Rogers, T.: *Fullest Commercial Use of Space: How the United States Should Go About Achieving it*, 1995
10. National Academy of Public Administration, *Findings: Commercial Space Processing and Requirements Forum*, March 1996
11. U.S. Congress, *National Aeronautics and Space Act*, 1958 (amended in 1984)
12. White House, *National Space Policy*, 1996
13. NASA, *Implementation of the Agenda for Change*, May 1996
14. Congressional Budget Office memorandum, *Budgetary Treatment of NASA's Advance Commitments to Purchase Launch Services*, June 1995

Report on Panel Discussion 2:

Entrepreneurial Initiatives to Use the ISS for Profit

I. Gracnar, A. Lindskold, International Space University, Strasbourg Central Campus, Parc d'Innovation, Boulevard Gonthier d'Andernach, 67400 Illkirch-Graffenstaden, France

e-mail: gracnar@mss.isunet.edu, lindskold@mss.isunet.edu

Panel Chair: J. M. Cassanto, Instrumentation Technology Associates, Inc., USA

Panel Members:

A. Eddy, Canadian Space Agency, Canada
B. Harris, SPACEHAB, USA
W. Lork, DaimlerChyrsler Aerospace, Germany
J. Manber, Energia Ltd., USA
J. Richardson, Potomac Institute for Policy Studies, USA
J. von der Lippe, INTOSPACE GmbH, Germany

J. M. Cassanto guided the second panel discussion, which centered on commercialization issues for the ISS. The discussion dealt with how to promote space travel to the public, going around competition barriers put up by the partners, and forming a working group from different nations. The panel also commented on keeping Mir operational to learn about commercialization issues at an early stage, and on whether completely private ownership of the ISS was a good idea or not.

A member of the audience began by stating that most space companies today were either government contractors or in other fields of business as well. He was interested in advice on how he, as a businessman, could raise the public's awareness of space and promote space travel to the public. **J. von der Lippe** suggested that business people should request more flexible operations of the ISS. **J. Richardson** wanted to encourage business competitors to go into space. **W. Lork** was of the opinion that space operations should be privatized as much as possible; when the "government rules" there are problems. **B. Harris** acknowledged the problem of profitability. The space business today is looking for a "magic bullet", the first product from space that will be profitable and act as a pathfinder for other products. No such "bullet" has been found yet; protein crystallization is the most promising one at the moment. **J. Manber** urged that the public should be better educated about what we are doing in space, and that

G. Haskell and M. Rycroft (eds.), International Space Station, 97-99.
© 2000 *Kluwer Academic Publishers.*

we have to be realistic about what we are doing there. **A. Eddy** conveyed the opinion that businessmen should talk to other businessmen about the benefits of space, thereby convincing the industry that space is an area in which to do business. Finally, **J. M. Cassanto** re-emphasized the need for competition. He gave an example of a "magic bullet" in the form of the encapsulation of drugs, and the use of the technique in other fields. He also saw further education on space, and why microgravity is important, as being necessary.

The next question concerned competition on the ISS and going around the barriers to that are put up by the partners of the ISS, e.g. by using hatches of different sizes. But named hatches could be used for advertizing. **J. Manber** said that the ISS was planned to be more commercial from the beginning. **J. von der Lippe** pointed out that NASA is the majority owner of the ISS and that we should be realistic about who has the "power" on the ISS. **J. M. Cassanto** was of the opinion that, since the ISS is international, it should be run like the UN, but without the bureaucracy; NASA does not know everything and should not dictate to ISS Partners. A UN type commission is needed to handle the commercialization aspects.

The panel was then asked how they felt about creating a working group, of different nations, on ISS commercialization. **A. Eddy** said that there has to be a balance between competition and cooperation. ISS is not a competition between governments, but a competition between private sector companies. We need cooperation between the governments so that the industries can work competitively. **J. M. Cassanto** referred to NASA's commercialization plan for the ISS as a "good start". The NGO must be an international consortium, containing governments, large industries and innovative small businesses.

A member of the audience was of the opinion that, if Mir were kept operating for a few more years on a commercial basis, then all the questions on ISS pricing policies would have to be solved in a very short time. **B. Harris** pointed out that, though the idea was certainly viable, the future of Mir is highly uncertain. However, if there is a private alternative to the ISS, it will force a lot of changes. **J. Manber** was very clear in stating that the commercialization of the ISS is "killing" Mir — an irony. Serious financial institutes have looked at Mir, but since they have felt that NASA would be a competitor they have decided to stay out of investing in the Russian Space Station. So, the fact that the ISS will be commercial stops investors from coming to Mir. **J. von der Lippe** urged everyone to be realistic. There is no market for the Mir; no industry is waiting to get aboard. **J. M. Cassanto** concluded by saying that, because Mir works, it should not be deorbited; it could be a "backup" for the ISS. NASA should speed up its commercialization plan. Users do not want to wait for 2-3 years to get their experiments into space. Since medical people cannot get a flight every 6 months they can lose interest altogether in going to space.

The final question concerned the concept of privatizing the ownership of the entire ISS. Boeing could, for instance, own the US component. Would such a scheme work? **W. Lork** replied that the main customer of the ISS is still the government. Experience with the Eureka platform showed that, once it had passed into private ownership, ESA was not interested in using it again because it was not "theirs". If the ISS is owned by the private sector, the space agencies will no longer be interested in supporting it any more. The problem is, in other words, that there is an insufficient number of customers. **B. Harris** thought it would be great if the governments only had a regulatory role on the ISS. Then the industrial partners would have to work together. **J. Manber** did not see the point of turning the ownership over to an aerospace contractor. A better thing would be to turn the ISS into a space hotel, a "Hilton in the sky". **A. Eddy** disagreed, saying that if the industry wanted such a hotel, it should build one. He did not see that things would change very much, since Boeing already runs the ISS. **J. Richardson** emphasized that businesses can fail; they do not provide guaranteed success. Also, there is a whole spectrum of ownership types between federal and private. **J. M. Cassanto** did not think the ownership should go to a big company like Boeing, since that would mean a conflict of interests. Rather, it should be given to an international consortium.

Session 3

ISS for Education and Public Awareness

Session Chair:

N. Ochanda, University of Nairobi, Kenya

Searching for New Opportunities from the International Space Station, and Using Them in Eastern Africa

N. Ochanda, University of Nairobi, Geography Dept., P.O Box 30197, Nairobi, Kenya

e-mail: ochanda@hotmail.com

Abstract
Universal acceptance of new opportunities from the International Space Station depends upon the meanings that such new opportunities have for members of the global family. While some members in the family are able to connect to the future and clearly see the value of the Space Station, others are only remotely aware of the new possibilities or they are just indifferent. Such people may not make reasonable choices or set goals when their attention is drawn to new opportunities. Instead, they desire the opportunities to fit in with their lifestyles; if not, the individuals become anxious and helpless and cannot conform to the requirements of the International Space Station and the global family. There are philosophical and political views that shape individuals' commitments to the ISS, their ability to construct meaningful lives and opportunities for moral and social change.

We aim at pointing out possible ways of making people who are far removed from the activities of the ISS to join the search for new opportunities. We believe that it is easier for individuals to join the search if they discover opportunities which they fit into their daily life occupations and if we can forge a link between converging Digital Technology, the Space Station and daily life occupations. We think that this should make the families working and living in Eastern Africa share the costs and risks, and reshape their labour and ingenuity in the search, along with personal achievement, family growth, social networking and spiritual nourishment. We are convinced that families can put their weight behind the Space Station and search for opportunities that enhance the lives of present and future global families.

1. The Search for New Opportunities

The International Space Station (ISS) has the potential to link individuals and families working and living in Eastern Africa to space ventures and services. When there is an opportunity to utilize it effectively countries, which are not directly contributing to ISS, would cooperate with ISS partners in international space projects. Individuals who are connected to the future reach the next space marketplace, fabricate the existing opportunities and create new ones from their daily life occupations. Much of the linkage would occur in the context of occupation that is stimulated by the meaning which new opportunities have for the individuals. The individuals would seek connectivity to their families and try to share intrinsically connected opportunities for family growth. Families would search for opportunities that link them to ISS network. Willingness to join the search would depend upon their experience to link with opportunities from the ISS and other areas of their daily life occupations. The individuals would pursue choices that bring about emotional, intellectual and

G. Haskell and M. Rycroft (eds.), International Space Station, 103-110.

social change [Reference 1]. These changes increase our chances of linking individuals and families to the ISS community.

1.1   International Cooperation in ISS Projects

Rationale for International Cooperation. There is the commitment by ISS partner states to utilize the capabilities of the ISS to the fullest, with the rationale being for long term internationally cooperative projects.   The ISS partners would promote international cooperation, perhaps using foreign policy mechanisms [Reference 2] to influence commercial utilization and operation of the ISS.   When the policy mechanism is applied within an atmosphere that stresses the value of ISS opportunities, the countries in Eastern Africa would be encouraged to cooperate in ISS projects.   Individuals who are connected to the ISS community may be sufficiently aroused to join the search and to build capability to fabricate a new life with new opportunities.   But when the policy is used differently, the dominant partners may be tempted to sideline "good will" propagation of the ISS and the countries would feel powerless and reluctant to cooperate in the international project.   Some individuals could become indifferent to ISS activities while those who are only remotely aware of them wait for privileges.   Potential users of the ISS in Eastern Africa may find it difficult to join the ISS community in the search for valuable opportunities from space when policy arrangements appear not to increase their participation in ISS activities.

Connectivity with ISS Partner States. Individuals, families and the social network in Eastern Africa should be connected to ISS utilization and operation through a network of an ISS community of users. In this way the views of ISS partners would increase connectivity and stimulate cooperation with non-contributing countries. The opportunity to link effectively is provided when the ISS partners foster interest among the non-contributors to perceive and deal with ISS suggestions.   Individuals would be motivated to search for opportunities with the conviction of value without losing their commitment to ordinary everyday activities and the interpersonal contexts in which they occur. Engagement in the ISS would provide an opportunity for commitment to justice and empowerment.   Connectivity would make individuals share new opportunities in a network. By communicating between themselves and with the ISS community, they would hopefully remove certain geographical barriers to ISS utilization and speed up its operations.

1.2   Control of Commercial Ventures

Commercial Ventures and Services. The ISS partners view the ISS as potentially the next space marketplace and are searching for ways of commercializing it. The rationale for commercializing the ISS would be the optimization of its potential, support for its growth and opportunities for private, institutional and public good. The search for new opportunities would be influenced by the control mechanism. If partner states take control of space ventures they would move between competitive and cooperative support for a diversity and growth of new ventures.

Countries in Eastern Africa could make initiatives, pay attention to new space ventures and take responsibility for sharing benefits, costs and risks. However, individuals would have little involvement and would, most likely, not be aware of space ventures. If the ventures were under the control of private sector this would provide new opportunities to individuals and families to form a network of space ventures and services that support ISS growth and a meaningful life.

Experience to Pursue Space Ventures. When ISS opportunities are brought to the attention of potential users, individuals with experience would reach the next space marketplace, extract new commercial opportunities and investigate opportunities in other areas of their daily life occupations. The individuals would seek connectivity to family needs. The families would cooperate as they engage in opportunities that link them to the network of the ISS community. Those without experience would feel left out and wait for privileges, as they desire that new opportunities fit into their daily lifestyles. They would become anxious and helpless if they cannot achieve personal growth and social worth. Institutions and the private sector lack motivation and commitment to search for new ventures and services. A large community of potential users in Eastern Africa does not have the experience or ability to search for or create other opportunities for their daily life occupations. Those who have the experience may lack the opportunity to search effectively.

2. Using ISS Opportunities in Eastern Africa

Using ISS opportunities involves a relationship between new opportunities, digital communications and other areas of daily life, and linking the results to the long-term political and philosophical views that stimulate engagement and shape the individual and family commitment to new opportunities. This helps to construct a meaningful life for individuals and families in Eastern Africa and to support growth and progress of the ISS.

2.1 Philosophical Views

The ISS cannot be separated from individuals' perception of the environment and the global network of users; we are likely to have to rely on this as the adaptive mechanism in the next century. If we use the ISS simply as a coping mechanism, some individuals would use the ISS for productive, touristic and festive occupations while others remain passive recipients. If we perceive the global village as a materialistic world, we can easily turn the ISS into an instrument for power and the movement towards commercialization and privatization suits the needs which we have in commercial terms [Reference 3]. Some individuals may use old approaches to new life experiences with the ISS, and rely on old skills. Used in this way, emotional states are stressed. A feasible approach is to incorporate ISS opportunities into our daily life experiences, which allows us to attach meaning to them. Our spiritual activities become integrated into ISS utilization as we retool old skills, and reshape dreams and capacities, to utilize the ISS. Adaptation becomes a process of selecting and organizing new opportunities to improve life opportunities on Earth according to the experiences of individuals. The outcomes are building a portfolio of investments and establishing strong ties with family and friends, and polishing our capacity to become a member of the global family here on Earth and later in outer space.

2.2 Political Views

Empowerment. We aim at helping individuals and families in Eastern Africa to redefine their position and alter their social network to suit their engagement in ISS utilization. Organizing and experiencing power determines the actual possibility of allowing ISS users to increase their potential to rise and move towards the ISS [Reference 3]. Empowerment occurs when individuals occupy decision-making positions in the ISS community. That captures other people's imagination and changes the terms of how Africans relate to the next space marketplace. This involves being open to the value of space realities. It can aggregate up to the national level, and increase engagement and commitment as the community fosters an interest in ISS activities. The individuals have to cope with tensions between the ISS community and their perception of opportunities fostered by the ISS community. The professionals feel motivated, change their attitude and become more open.

Emancipation. Engagement in the ISS programme should be done in such a way that creates a fair opportunity for everyone, regardless of ability and ethnic differences (Reference 4). This could involve tapping creative, pragmatic and utilitarian potentials. Sharing of opportunities becomes an everyday experience of power in exercising choice and self-control in a social network.

2.3 Reaching the Next Space Marketplace

State Interest. Countries in Eastern Africa are strengthening their positions, in horizontal power structures with progressive leadership, and are accorded positive per capita income. Many are improving their communications systems for local and global sharing. This forms the context in which Eastern Africa could be more involved in ISS utilization and operations. By turning ISS opportunities to their own ends, countries of Eastern Africa may shift some of their research and educational needs; entrepreneurial and cultural activities can be turned away from traditional approaches towards ISS opportunities. New opportunities in the next space marketplace would increases state revenue as they improve the lives of their citizens, caring and sharing opportunities with other countries.

Individual and Family Interests. When we use the ISS in the context of daily life, individuals and families would reach the next space marketplace and demonstrate creativity and enterprise. Given power and opportunity to make reasonable choices, the individuals would fabricate a new life from existing ISS opportunities. New feelings and ideas could be evoked as the desire in individuals to choose opportunities that meet household needs is increased and families would share intrinsically connected opportunities in a social network. With ISS opportunities, individuals would create new opportunities from interactions with daily life activities drawing on threads of their past selves and creating a new routine. We should, therefore, identify ventures that are effective in meeting local needs.

2.4 Socio-Spatial Network of Customers

Local Control. Daily participation in the ISS accrues; it is culture laden and is connected to the societies in which we live. It is our moral identity that is likely to ignite intense engagement in the ISS. This requires a policy environment that is hospitable to access to ISS opportunities and human talent. Individuals would be motivated to identify new space ventures, and an emotional intensity would be involved when the individuals forge a reflective process of making choices and constructing daily routines with ISS opportunities. This would require a knowledge of ISS opportunities suitable for daily life, of how engagement in them shapes personal identity, and of how to use reflective process in order to enhance life opportunities. We enact social roles that maintain the stability of our culture, which are tied to moral identities that make us feel passionate. Occupational roles involving ISS fall within social roles and let us deal with a full round of daily activities. We must ignite an intense engagement in ISS opportunities and daily life activities, and feel passionate about those activities that preserve or enhance our moral identity.

Social Network. The most important component in social networking is the connectivity of communities in Eastern Africa and the ISS community, and the capacity to understand and use the connectivity to gain access to ISS opportunities. Tele-centers could be the main delivery mechanism to facilitate such a connectivity by generating income activities and developing a market for ISS ventures and services.

National and Regional Strategies. Each nation could facilitate community access to ISS through a policy mechanism involving human capacity and community needs. A policy that links university and other networks leads to a sharing of ISS opportunities, research on issues central to the introduction of ISS opportunities and national information, and dynamic partnerships that apply throughout the region. Some of the ISS ventures and services may cut across national borders; tele-centers can facilitate community usage of the ISS, human resource sharing and sharing the lessons learned.

3. Moral and Social Change

3.1 Motivation to Enter the Next Space Marketplace

We should be motivated to reach for the ISS; we should develop the ability and willingness to search for new opportunities that we think are useful to the countries in East Africa or those that families expect from individuals. We explore ourselves through the real experience with the ISS, and our emotions link us to ISS opportunities; cognitive abilities make us aware of our abilities to search for new opportunities (Reference 4). We do this as we conform to social expectations and comply with expectations that are appropriate to our specific tasks. These depend upon the ways in which we use the skills we possess as participatory members, the ways in which we organize ourselves as we communicate with others, and our interests. We succeed when we apply our searching and use our skills in teamwork, share our experiences with others and establish companionship in our families. The self awareness that is so intrinsically linked to society is an opportunity to learn to celebrate our talent in the utilization of the ISS for quality living to uphold fairness and equity — or it can be a barrier if we learn to be impaired in the everyday situation. We break bad habits and avoid the unthinking, patterned and rigid way of life when grasping the new ISS opportunities, and educate our children about space education.

3.2 Socio-Spatial Network of Users

Our experiences with the ISS and other patterns of thought locate us in a physical world. We forge new time and space configurations by breaking geographical barriers in the utilization and operation of the ISS [Reference 5]. In the ISS community, individuals flourish and create order, which they find personally gratifying; they proceed according to the demands placed upon the time. Individuals in nations, which are not directly involved in the ISS, have their potential limited by inexperience. We forge a new use of space and time by creating a socio-spatial network that connects the international community of ISS users.

3.3 Valuing ISS Opportunities

We can engage in ISS utilization and operation with the conviction of value that contributes to the next space market. We must neither lose our commitment to ordinary everyday activities nor to the interpersonal contacts; we must value the customer in Eastern Africa. Value is built during life experiences that evoke strong emotions. If value is built through association with past life situations a change may be sparked as we engage in ritualized activities of merging opportunities with our daily life occupations. If value is built in relation to the larger human condition, social change occurs as we fully participate in the ISS community. Valued opportunities are a powerful means of companionship in families and lead to the formation of a social network.

3.4 Personal Achievement and Family Growth

Individuals with visions of hope investigate other areas of their daily life occupation, create new opportunities and celebrate personal achievements. Those without hope risk being left behind, and yet this should not happen. In societies in which visionary ideas are pursued, families engage and co-operate in occupations that link them to a social network as they celebrate their growth. Societies without a vision are tempted to turn ISS opportunities into instruments of power and this may tend to suppress personal achievement and family growth. The best way to foster interest in the ISS opportunities is to participate. Non-space communities can get interested and involved in ISS activities.

4. Reflections and Conclusions

• **Reflections.** What is the policy framework that is hospitable to ISS access for countries in Eastern Africa? What contribution could these countries make to ISS operations? What are the different approaches to providing access to ISS ventures and services? And how can they be used to extend the reach of the individuals and families working and living in Eastern Africa? How could the individuals and families identify ventures and services that are efficient in responding to personal achievements and family growth? And how can they contribute to the growth of the ISS?

- **Conclusions**. Non-contributing countries could become more involved when ISS partner states find value in their contribution and foster their interest in such an internationally cooperative project. Individuals, the private sector and institutions in non-space communities will be committed to ISS utilization when they discover ventures and services that fit in their daily life occupations. Connectivity with the ISS is faster and more effective when directly made between individuals, the private sector and institutions, rather than via policy mechanisms.

Acknowledgements

I am most grateful to the UN Office for Outer Space Affairs and the International Space University for sponsoring my participation in the international exchanges and insights of the next space marketplace. I acknowledge the University of Nairobi for allowing me to join the space culture during this symposium on the International Space Station.

References

1. Meredith, C.: *Goodbye Gaia, goodbye!* Paper presented at the 14th Annual Conference of Science Teachers Association of Western Australia, Muresh, 1991
2. Le Gall, Jean-Ives.: CNES - The way forward: A new strategic plan. *Space Policy, 13 (1)*: 1536, 1997
3. Homer-Dixon, T.: The ingenuity gap: Can poor countries adapt to resource scarcities? *Population and Development Review, 21 (3)*: pp. 587-612, 1995
4. Townsend, E.: Occupation: Potential for personal and social transformation. *Journal of Occupational Science, 4 (1)*: pp. 18-26, 1997
5. Dear, M.: Time, space and the geography of everyday life of people who are homeless. Occupational Science: The evolving discipline, R. Zemke and F. Clark (Eds). Philadelphia: F.A. Davies, 1996

International Space Station: New Uses in Marketplace of Ideas

J.D. Burke, 165 Olivera Lane, Sierra Madre, CA 91024, USA

e-mail: jdburke@its.caltech.edu

Abstract
Education and public outreach are recognized functions of the International Space Station (ISS). In this paper let us consider how those functions may be enhanced as ISS becomes fully operational and can support new uses and new users. We shall focus on the International Space University (ISU) as the primary vehicle for a future advanced education program using ISS, extending the achievements of present outreach efforts. In 1987 the founders of ISU visualized a three-step process: first, peripatetic ten-week summer sessions; second, a year-round Master of Space Studies curriculum at a central campus; and third, a campus off-Earth. Their goal was to build a worldwide network of leaders whose shared, intense educational experience would raise lasting friendships. The first two steps have now been splendidly achieved, with 1350 ISU alumni already making their mark in space enterprises. Not only was ISU an idea whose time had come; the felicitous growth of the Internet greatly aided it in building an intercultural academic community that now includes 25 affiliate universities in 14 countries. Arthur C. Clarke, ISU's Chancellor, regards ISU as a modern analogue of the great universities of the renaissance and enlightenment, when transocean voyaging coincided with an outburst of new institutions and ideas. Permanently occupied space stations in low Earth orbit can be thought of as the coastal trading vessels of a future where people regard ventures farther into space as a practical reality. The commercial success of space fiction entertainments shows that the public is ready to believe in such a future.

However, for the notions of space tourism, revived lunar and martian exploration and lunar settlement to become reality, there must be public acceptance of a large and sustained investment in real as distinct from fictional off-Earth living. Use of off-Earth resources is essential. In such a large shift of public opinion education will be the key; self-serving agency propaganda will not do. A broad segment of the world public needs to share true knowledge and a rational belief about the long-term value of human space voyaging and, more fundamentally, of the elevation of human values that can accompany it. Existing commitments to education and public outreach provide a model for the early stages. Once the ISS becomes permanently occupied a more diverse and unpredictable education activity can start to grow. Diverse, because more educational experiments will be possible; unpredictable, because people both on Earth and in orbit will invent new teaching and learning concepts. A worldwide information system already exists and the first experiments are in progress for using it to support spaceborne education. However, many questions as to policy, economics and above all content remain unanswered.

ISU is the right institution and this symposium is the right venue for starting the needed discussion. In this paper we examine possible paths of evolution from today's programs and plans to a time when there is in space an educational, research and public-service enterprise strong and durable enough to be called an affiliate campus of the International Space University.

1. Introduction

A founding principle of the International Space University is that humans will eventually establish permanent settlements off the Earth. Though no one can now say when, with what purposes, or driven by what incentives that

G. Haskell and M. Rycroft (eds.), International Space Station, 111-118.

development will occur, we already need to begin considering its implications and nurturing its future leadership. The history of exploration and colonization on Earth includes both sublime and abysmal examples of human behavior. In space there is a chance for humanity not only to repeat past mistakes and make new ones but also to reach new heights of courage, achievement, knowledge and wisdom.

Space stations in low Earth orbit offer the first technical capacity for sustained living off-Earth. Heroic feats of endurance in Mir demonstrate the remarkable resilience of trained and motivated humans. In the next step, ISS is intended to provide augmented living leading to increased human performance and increased knowledge of the interaction of humans with the space environment. These complicated international programs depend absolutely on public support and the organized skills of large numbers of professionals highly educated in many fields. So too will an even more demanding future where people inhabit the Moon and venture to Mars.

Recognizing this, people in many nations are promoting space-derived education and public outreach as stimuli not only for future space achievements but also for the health of their economies. That is good, but there is more to the subject. In this paper let us entertain the thought that ISS, in its later evolution, may play host not only to the scientific, technical and managerial education required for programs and economies but also to the liberal arts, elevating public understanding and commitment to an open, sustainable future, thus advancing humanity toward the goals of the founders of ISU.

2. Present Space-derived Education and Outreach

Let us begin with a brief survey of space-related education in the United States, the country whose education establishment has been most impacted by the coming of the space age. For more than a century all Americans have been expected to go to school. Over much of that time, pro and con arguments over education policies flourished even as most of the population became literate and economically functional and a few institutions achieved world renown. But all was not well in the resulting education structure. The great research universities, sponsored by government after the technical miracles that helped win World War II, turned out a generation of scientists who won many Nobel Prizes. Meanwhile primary and secondary schools were failing as shown by test results and, more important, by an observable growth of scientific illiteracy and defective thinking among non-scientists [Reference 1].

Upon the launch of Sputnik in 1957 a perceived crisis was followed by efforts to improve American educational performance. Space-related education and outreach were regarded as significant contributors, especially in encouraging students to pursue math and science. Today in the United States a broad and diversified effort at national and local levels attempts to advance education in all school and university grades. Success varies from place to place but educators observe that, for many reasons, needed change comes slowly [Reference 2]. In an effort to hasten progress, NASA and other agencies are augmenting their education and outreach activities. A typical example is the recently reorganized Web site for NASA's human flight enterprise [Reference 3] giving information about a variety of school and university programs bringing students and teachers into closer contact with workers at the NASA Johnson Space Center. Among those programs are several intended to make use of ISS capabilities; precursor experiments are already in progress using shuttle flights [Reference 4].

In addition to initiatives by NASA and the National Science Foundation, many other public and private space-related educational enterprises exist in the US. By visiting some typical Web sites [Reference 5] one can gain an idea of the effort being thrown into educating children, teachers, university communities and the general public, including emphasis on women and minorities who have been under-represented in fields related to spaceflight. At the university and postgraduate level the space science enterprise is threatened by the combination of events outlined in Reference 1. The collapse of the Soviet Union, combined with the over-production of PhD's in postwar decades, is foreclosing opportunities and driving bright young Americans to seek more rewarding careers away from science. Graduate students from other countries, however, still see opportunities and are now doing much to energize US universities.

What does all this mean with regard to ISS? As in every other NASA program there is an education and public outreach component [Reference 6], a source of funding and encouragement for building bridges between ISS personnel and the outside world of educators and the public. But the ISS is much more than a usual NASA program because of the large and critically-important contributions of America's international partners. The full educational implications of this difference remain to be seen, but one obvious consequence is that ISS outreach efforts will need to be adapted to existing structures in the partner nations. And, beyond that, a goal of ISS is to be a resource of knowledge and inspiration for the entire world.

Recognizing these new challenges, NASA staff and their colleagues in the partner nations are preparing a suite of ISS-based education and outreach activities [Reference 7] using the Internet and other means to engage as large an international population as possible during the coming years. This ISS outreach can be the springboard for the enterprise proposed in this paper, namely, the evolution of ISS into a test-bed precursor of schools and universities off-Earth.

3. The Long-Term Promise of ISS

To discuss our subject meaningfully we must make three main assumptions: first, that ISS will succeed technically; second, that its partners will stay the course and continue to operate it after its initial goals have been achieved; and third, that humans will ultimately use space resources and live in large numbers off-Earth. Of course the third is the most speculative concept. However, without that as a clear goal, even though its achievement may lie in the indefinite future, much of the justification for a long-sustained ISS program disappears. Once the problems of long-duration human spaceflight are worked out well enough to support a rational commitment to the human exploration of Mars, why will we need to continue ISS operations? It is easy to predict a budget collision between that continuation and the Mars program itself. Perhaps commercial opportunities can provide a strong impetus as discussed elsewhere in this symposium, but as yet that proposition is not proven.

Resolving this conflict will require the development of good and accepted reasons, both commercial and non-commercial, for continuing ISS operation into an extended mission phase. In robotic space projects extended missions are common. Magellan, having finished its radar mapping of Venus, was maneuvered into a low orbit (using a tiny amount of residual project funding) and tracked to give a gravity map of the planet. The two Voyagers, having long since achieved all of their planned objectives, are operated at relatively low cost as heliospheric probes traveling into interstellar space. Galileo, having met its main scientific goals in the Jovian system despite a failure of its high-gain antenna, has been funded to continue operations with emphasis on Jupiter's mysterious satellite Europa.

Each of the projects mentioned included an education and outreach component, but the main purpose of their extended missions was science, not education. In contrast, operation of the NASA Deep Space Network (DSN), a worldwide complex of tracking and data acquisition facilities, includes one highly-successful activity entirely dedicated to education. This is the Goldstone-Apple Valley Radio Telescope (GAVRT) project in California [Reference 8]. In this project a 34-meter tracking antenna, decommissioned from service in the DSN, has been converted to serve as a radio telescope and is made available to middle school and high school students who design experiments, make observations and report results, operating through the Internet. Funding for telescope operation and maintenance is provided by NASA under a memorandum of understanding with the Lewis Center for Educational

Research, a public-service institution set up to serve communities near the Goldstone facility. Because of its Internet communications, however, GAVRT is not limited to use by local schools; distant student teams can and do use it. The whole activity is highly regarded as an example of innovative sponsorship, education and outreach benefiting from a previous large investment in the space program.

It is not a huge extrapolation to imagine a late-term ISS activity analogous to GAVRT. Education and outreach, at least as they are defined today, will not command such large budgets as to cover the whole cost of maintaining a Space Station, even in a low-activity, extended-mission mode. However, by innovatively combining commercial and public sources of support it may be possible to sustain the ISS and use it in a new fashion — to make it serve as a marketplace of ideas. Let us now examine this prospect.

4. Steps toward a Campus in Space

Once ISS assembly is complete and operation passes from preparatory to routine status, much of the on-board timeline will be devoted to activities intended to generate new knowledge. Opinions differ as to the efficiency of these activities, but if assembly is successful it is only a matter of time until experience will show what manifested programs tend to have the best yield in proportion to the resources (funding, transport, microgravity, IVA, EVA) that they demand. Based on prior experience in the management of timelines, such as that of the Hubble Space Telescope, against multiple demands it is to be expected that understanding how to manage the ISS resource will evolve and improve over a period of years.

During this evolution some part of the ISS resource will be devoted to education and public outreach as outlined in Reference 6. However, progress may be limited by the rate at which schools can adapt to the new information exchanges made possible by communications with both human and robotic ISS systems. Much will depend on the individual initiative of students, teachers, school boards and administrators.

One interesting question is the relation between ISS-based education and space-supported education delivered by other means, such as small store-and-forward satellites typified by the Surrey UOsat series [Reference 9] and the student-built satellites launched from Mir. ISS educational exercises conducted by crew members can offer a real-time human element just as immediate as that of a classroom, if and only if the students on Earth are equipped with appropriate video terminals. Many US schools now have such facilities and use them for receiving NASA television. In some school systems optical fiber networks permit two-way TV interactions with multiple remote classrooms. A lower-cost alternative is Internet slow-scan imaging through ordinary telephone lines. However, even this technique is as yet unavailable in parts of the developing world; it should be a prime objective of space-supported education to reach such parts. For example, Brazil, a modern nation with highly-developed industry and communications in urban centers, is expected by year 2000 to have only 12 telephone lines per 100 of its 175 million people [Reference 10]. Bringing education of any kind, space-delivered or not, to such populations remains a challenge, but numerous entrepreneurial enterprises are trying to meet the need. If they succeed it will be imperative for the ISS education and outreach component to use the resulting new capabilities.

More important than these technical options is the question of content. Even within the US it is a continual struggle to establish content appropriate to the learning capacity of students at all ages. Educators sponsored by the National Academy of Sciences have established some minimum, voluntary standards [Reference 11]. Whether or not these would be appropriate in other cultures is an open question. Yet if ISS is to fulfill its promise as a beacon of knowledge and learning incentives its education standards must somehow fit world needs, not only in terms of what reaches students but also in the professional development of teachers in many countries. This immediately raises the question of language barriers. Clearly a future ISS education plan should deal with that problem to the extent possible. There is some useful experience in the comparison of children's performance in different countries [Reference 12]. ISS education plans would benefit from an extension of such data-gathering and analysis.

From all of these observations it is apparent that ISS education and outreach will demand a multicultural approach. That is the main reason for what this paper advocates — namely, the accession of the International Space University into a significant role. Within the US there are good models for such a function. One example is the Space Science Institute in Boulder, Colorado [Reference 2]. This organization acts to bring together educators, space project managers, scientists, museum directors and interested citizens so as to catalyze improved space science education at all levels. It is highly successful in its chosen field. However, that field represents only a part of what should be contained in the ISS education and outreach enterprise. For ISS to reach its full potential as a world resource it must offer not only science but also learning in all the other fields where space-delivered education can make a difference.

This provides another reason for urging that ISU become an agent: ISU from its beginning has offered an interdisciplinary curriculum intended to broaden the viewpoint of its alumni and fit them for leadership in a world where engineering and science interact with policy, law, business, history, politics and the arts. If ISU becomes engaged in the development of a worldwide ISS outreach integrating these many aspects of civilization, ISU will do much to enhance the benefits of ISS to humanity. A practical way to start along the recommended path would be for ISU students to carry out a project on ISS education and outreach pointed toward use of space resources and human space settlement. The current Team Project on ISS commercialization [Reference 13] includes an ISS users' guide, excellent grounding for such a follow-on ISU project.

5. Conclusion

During the years of its operation the ISS may support a variety of noncommercial, educational and cultural activities. At first these will inevitably be constrained, not only by on-board limitations but also by the limits of ground facilities. It is unrealistic to imagine that every school in every country can gain early access to ISS information. However, over the years of growth of ISS and its evolution into a world asset, economic and business drivers will be forcing the proliferation of universal communications systems. In time it may indeed be practical to hook the ISS into an existing worldwide information grid enabling students anywhere to communicate with people and robots aboard the station. Then the limitation will not be technical; instead it will be due to the same factors that now, as outlined above, impede educational change. Long before that stage is reached, ISS-based educational innovations should be in pilot testing;. what the results may be, we cannot now tell. But we can observe successful analogues. The Goldstone-Apple Valley Telescope (GAVRT) consortium, the Space Science Institute at Boulder, and other innovative efforts are showing the way. Going beyond science education, there may be an opportunity to propagate the ideas espoused by G.K. O'Neill and promoted by the Space Studies Institute at Princeton [Reference 14] — that space manufacturing using extraterrestrial resources can be the key to an open and expanding future for humanity. Only through the widespread acceptance of those ideas is broad public support likely for the concept of large human settlements, including schools and university campuses, off-Earth.

Public support of a space-enabled better future need not be limited to the present spacefaring nations; indeed it must not be. Improving life in the developing world is an urgent goal for all of education, including that supported by ISS. Ideas promoted in the World Academy of Art and Science [Reference 15] are typical of current thinking on this problem. From its beginning ISU has included people from developing cultures in every class of students. If it proves to be possible, through some combination of commercial and other incentives, to sustain an extended mission of ISS there will be an opportunity for the station to serve as a catalyst in a new marketplace of ideas. By demonstrating the practicality of off-Earth living plus a commitment to education in the values of a learned, civil terrestrial society encouraging all the arts and sciences, people in the station and their colleagues on Earth may open the way for humanity into the cosmos. The International Space University should aim to be a primary agent of this outreach.

Acknowledgements

The author wishes to thank Ms. Lindsay Chestnutt and Prof. Michael Rycroft for review and editing. Ms. Goldie Eckl, with ISU since its founding, inspires all to sustain their commitment to ISU's goals. Dr. and Mrs. Max Larin, Dr. Knut Oxnevad, Mr. Clovis de Matos and Ms. Olga Zhdanovich have provided ideas and stimulating discussions. Max, Knut, Clovis and Olga are alumni of ISU and each is fulfilling ISU's promise by doing good work in today's space enterprise while maintaining a thoughtful interest in the farther future.

References

1. Goodstein, D.L.: *Scientific Elites and Scientific Illiterates*, Engineering and Science, Spring 1993, 23 - 31.
2. Morrow, C. et al.: *Fifth Annual K-12 Education Workshop for Scientists, Engineers and EPO Leads*, Boulder, Colorado, 11 – 14 April 1999
3. http://www.spaceflight.nasa.gov/outreach/index.html
4. http://www.calspace.ucsd.edu/ed-outreach.html
5. http://www.nas.edu/rise, http://www.nsip.net, http://learners.gsfc.nasa.gov, http://www.nsf.gov/sbe/srs/women/start.htm, http://cass.jsc.nasa.gov
6. http://www.station.nsa.gov/outreach/
7. http://estec.esa.nl/outreach/
8. http://deepspace.jpl.nasa.gov/dsn/applevalley
9. http://www.ee.surrey.ac.uk/
10. *Globalstar Annual Review 1998*:   http://www.globalstar.com
11. *National Science Education Standards*, U.S. National Research Council, 1996, Washington, DC: National Academy Press
12. http://www.csteep.bc.edu/timss
13. *International Space Station Users' Overview*, Team Project of the ISU Master of Space Studies Class of 1998/99
14. *Space Manufacturing 11* : Proceedings of the Thirteenth SSI/Princeton Conference, edited by B. Faughnan, Space Studies Institute, 1997
15. http://www.hhh.umn.edu/centers/world-academy/page5.htm

ISS — The First International Space Classroom: International Cooperation in Hands-on Space Education

V. A. Cassanto, ITA, Inc., 15 E. Uwchlan Avenue, Suite 408, Exton, PA 19341, USA

e-mail: Vcassanto@aol.com,   Hyperlink: http://www.ITAspace.com

D. C. Lobão, Instituto de Aeronàutica e Espaço IAE/ASA-L, Praça Marechal Eduardo Gomes n. 50, CEP:12228-901 São José dos Campos, SP Brazil

e-mail: Dclnaee@iconet.com.br

Abstract
The present paper explores the possibilities for international cooperation through hands-on space education onboard the International Space Station. The purpose is not only to provide a direct and real-life educational experience but also to link together students and teachers from many countries as an integrated team with a common goal. The role of the International Space University is important in facilitating these student international cooperation efforts on the International Space Station.

1. Introduction

The International Space Station has a multitude of exciting opportunities for space education relative for both students and teachers. Real opportunities for student space research on the ISS will help to foster the education of our next generation of teachers, students, and space researchers and explorers. As a result of a private sector, international space education program with Brazil, France and the United States, using the Space Shuttle for student space experiment opportunities, this paper shows how a similar system can be utilized on the ISS. Ways are explored to enhance the "hands-on" aspects of the ISS experience, for example, multi-national multi-school space experiments, student down-link visualization of the experiments operating on the International Space Station, an international e-mail network among students and teachers to discuss and plan experiments, student and teacher direct e-mail and/or video conference communication with ISS Payload Specialists, and the development of a unique program such as the proposed Student Space Gardens — growing vegetables for astronauts to consume while they are on the Space Station. The ISS will be a perfect follow-on and larger scale motivator for the current private sector hands-on space education work being conducted on the Space Shuttle.

Here we emphasize the use of the International Space Station by schools and universities as an integral part of international cooperation in space

G. Haskell and M. Rycroft (eds.), International Space Station, 119-125.
© 2000 *Kluwer Academic Publishers.*

education. The basic idea is to motivate and realize an educational opportunity for students and teachers involving space activities.

2. General Aspects of Space Education

Space education provides an ideal entree to the methods of science in general [Reference 1, 2]. When students begin to understand how it is possible to learn about space science, they obtain a demonstration of the power of the scientific method that is likely to have a beneficial impact in all their science courses. Many teachers would love to include more space education in their classes, but they are often held back because of their own backgrounds and training in the subject which may be weak or out of date. They worry that, without further training, they will not do a good job teaching space education subjects. Yet how many teachers have the time to spend a semester learning astronautics and astronomy?

That is the reality in many countries around the world. Most space educators prefer to give support and emphasize all kinds of hands-on activities. It is expected that students will discover the ideas of space or astronomy for themselves, not just to read or hear about them passively. Space education has two sides that work together: one is to train people who will later work in and for space activities of all kinds, and the other is to develop space awareness among the community, let people understand space activities, and get the community to consider and support space activities [Reference 3].

The well-trained teachers can return to their schools with the resources, activities, and information to incorporate more space education into their existing classes. So in order to have a successful cooperation in hands-on space education, basic work in education has to be developed in the schools; teachers have to be trained to be effective agents of learning and change in the classroom. The First International Space Classroom, aboard the International Space Station, can motivate students in the sciences and help them to develop their future careers.

3. Hands-on Space Education Program – an Example

A private space company, ITA, Inc. (Instrumentation Technology Associates) successfully established a student hands-on space education program, which has been running for the past eight years [Reference 1, 2]. Many student microgravity experiments have been flown on ITA's payloads (designated CMIX and CIBX) on the Space Shuttle (missions STS-52, STS-56, STS-67, STS-69, STS-80, STS-95) [Reference 4] and onboard the Brazilian VS-30 sounding rocket (March 1999) [Reference 5]. The program to date has involved over 3000 students and 50 teachers nationally and internationally.

3.1 The Space Experiment Equipment

The students' experiments were housed in an automated laboratory, the Materials Dispersion Apparatus, or MDA [Reference 1, 2]. The MDA is an automated device for materials processing in space, capable of bringing into contact as many as 100 different samples of fluids and/or solids at precisely timed intervals. The MDA provides ready-to-use equipment for students. ITA handles all of the NASA integration and documentation [Reference 2, 4], thus freeing the students and teachers to concentrate on the experiment design and preparations. The MDA units provide the students with an easy and flexible way to conduct a variety of experiments in space.

ITA has flown a variety of seed experiments for elementary and high school students, i.e., coreopsis, columbine, mustard-spinach, mushroom, tomato, pepper, poppy, radish, lettuce, and alfalfa seeds. The tomatos in space experiment is a joint educational adventure by ITA and Lockheed Martin involving young students from various U.S. elementary schools [Reference 1, 2, 6]. The objective of the experiments is to determine whether exposure of tomato seeds to different space environments — Shuttle middeck, Shuttle cargo bay, Long Duration Exposure Facility (LDEF) — and different exposure durations (from 9 days to more than 4 years) affects germination and growth rates.

3.2 The Program

An ITA representative goes to the school to teach, or the school's students and teacher come to ITA to learn about, microgravity, the experiments already performed, what is possible, and the equipment. Typically one teacher or one student interfaces with ITA to design the experiment and document the materials to be flown and the experimental protocol. The number of students involved in each school ranges from a few to many. In addition, for some experiments, several schools were involved, thereby spreading the experience to many students [Reference 1, 2].

In the case of high school students, the ITA space education program has had several beneficial effects on their career choice and education. Some of the students have expressed this positive influence in their lives. For instance, a student from the Titusville Florida High School reported that the experience had caused him to focus on science as a career. In general, the main benefit was that it motivated a new interest in the sciences and mathematics through their participation in a hands-on science project associated with the high-technology and high-visibility space program. They were involved in a space program with specific objectives, real progress milestones leading to a Space Shuttle flight, working in close coordination with faculty, scientists and engineers. They learned the value of searching libraries for relevant information, planning experiments, calculating parameters such as packaging density, preparing flight samples, obtaining experiment samples after the flight, evaluating the results, comparing with ground control experiments, and writing reports.

3.3 International Involvement

3.3.1 France

Following the success of the ITA student program at the U.S. national level, the same type of program was offered internationally. Through a collaboration with the International Space University (ISU), a competition was initiated for French high school students to fly an experiment on the STS-95 (John Glenn) mission. The ISU interdisciplinary and international approach was adopted by the French students, and this led to a more complete experience [Reference 7]. Experiment preparations included teamwork among all the different school departments (e.g., English classes writing the press release and the experiment report in English, and art classes designing the mission patch and commemorative card). The students received local corporate sponsorship to pay for their flights to the U.S. and lodging in Cape Canaveral to attend the launch. The French students had earlier made contact with fellow American students from the Travis Middle School in Texas. When they arrived in Cape Canaveral, they greatly enjoyed meeting each other and sharing in the pre-launch and launch activities [Reference 8].

3.3.2 Brazil

Following an important workshop in Brazil, "The First Brazilian Seminar on Space Education," in 1997 [Reference 9], both authors made efforts to include Brazilian Universities in the ITA program. On the STS-95 Shuttle mission, three Brazilian University experiments were flown [Reference 5]:

- Universidade do Vale do Paraiba (UniVap) University: to study the effects of microgravity on the regeneration of body parts of the Planaria worm
- Faculdade de Engenheria Industrial (FEI): to better understand Lipase enzymatic reactions in microgravity
- University of São Paulo: to obtain high quality sugar crystals from space.

The Planaria and Lipase experiments were re-flown for validation and further experimental data on a Brazilian (VS-30) sounding rocket providing several minutes of low gravity. An additional University of São Paulo experiment was included: "Crystallization of Pharmaceutical Products." The rocket mission named "Operaçao São Marcos" was organized and sponsored by the Centro Técnico Aeroespacial, and the Instituto de Aeronàutica e Espaço, of São José dos Campos, Brazil [Reference 10]. This was the first time that microgravity experiments were flown on a Brazilian sounding rocket, and they were done by teachers and students! These kinds of experiments are being encouraged by the Brazilian Space Agency as a stepping stone for utilization of the Brazilian share of the ISS.

4. Student Space Gardens

The proposed Student Space Gardens study provides a perfect follow-on from a multitude of student plant biology experiments already performed on the Space Shuttle [Reference 11, 12, 13]. It is also an excellent program for students of all ages (ranging from the very young, i.e., eight years old, to adults). Through hands-on activities, experiments, and discussions, students practice how to identify, classify, organize and recall information, which helps to reinforce their basic skills.

In long duration space flight (i.e., the International Space Station, future manned missions to Mars, etc.), plants will be required as a food source for the crew. A good example is the soybean. Soy is unusually complete in proteins: of the eight essential amino acids, soybeans contain seven in sufficient quantity, thus making it an important human food [Reference 14]. Plants, in general, can be an important part of bioregenerative life support systems in which food is produced, human waste is recycled, air is purified, carbon dioxide is consumed, and oxygen is manufactured.

The objectives of the experiment aboard the ISS would be to:

* find out what seeds can be germinated in space
* discover the optimal plant varieties that may be suitable in life support systems as well as food sources in space
* optimize seed production in space.

To start, students should build a small open garden to understand how to grow vegetables from seeds. Then the garden needs to be modified, miniaturized and enclosed (accommodated to existing flight hardware) in order to be flight-worthy for the ISS. The ground-based garden should be identical to the ISS garden in order to be the ground control.

To demonstrate the effects of microgravity on seeds, students set up the experiment, monitoring plant growth and appearance with the frequency of watering, water temperature, exposure to fresh air, soil, and light being as constant as possible. They select rapid-growing seeds as well as slow-growing ones.

5. Communications

The conventional e-mail facility has to be intensively used as the main means of communications for information exchange. A World Wide Web page should be held as a complement part of the monitoring process during the entire experiment period. This network will bring as much information as possible to satisfy all students' needs.

Reports of the experiment can be monitored via direct e-mail communication onboard the ISS. A special server can take the e-mails and provide the reports as quickly as possible to anyone participating in the experiment. This will keep all those involved up to date about the status of the seeds in the microgravity environment. In case of any changes, the ground control equipment can immediately be adjusted to replicate the conditions onboard the ISS.

A basic aspect of this First International Space Classroom is the visualization of the experiment operating on the International Space Station. This can be done by a down-link direct to one specific school or over a network reaching as many schools as possible. This effort presents a major advantage as a live operation that carries visual pictures plus sound.

6. The Role of ISU

The ISS will be comprised of modules from different partner countries, but currently there is no one module that is truly international to act as a reminder and a symbol of the importance of international cooperation and space education [Reference 7]. The ISU could propose a module which could provide a "uniting" function, being at once a true space educational facility, an orbiting space "library" and, perhaps, a common neutral meeting place aboard the ISS. That module could be the First International Space Classroom. We encourage the ISU to make this program on the ISS a reality for the next century.

7. Concluding Remarks

The overriding goal of this First International Space Classroom on the ISS is to motivate and teach students to become well-formed professionals, i.e., to view the space field as their professional arena, to be committed to contributing to it in the future, and to understand the technical, logistical, financial, and organizational issues that shape the space field. Since the space field is young, we expect our students to be educators who, regardless of their eventual professional settings, can speak knowledgeably about the space field and its role in our daily lives. We accordingly emphasize the development of skills in space multidisciplinary areas, with a set of hands-on experiences.

Acknowledgements
The second author would like to thank Miss Valerie A. Cassanto and Mr. John M. Cassanto, President of ITA, for their unwearying support in making space education a reality for students and teachers in many countries.

References
1. Cassanto, V.A., Wood, B.J.: *Student Experiments in Space: A Private Sector Model*, 48th International Astronautical Congress, Turin, Italy, October 6-10, 1997
2. Cassanto, V.A., Wood, B.J.: Student Experiments in Space: A Private Sector Model, *Earth Space Review, Vol. 7*, no. 2, pp. 25-28, 1998
3. Pignolet, R. G.: *Second Workshop on Aerospace in Portugal* (Workshop Sobre Actividades Aeronàuticas e Aerospaciais em Portugal), Instituto Tecnico-Centro de Congressos- Lisboa, July 29-30, 1998
4. Morrison, D.R, Cassanto, J.M.: Low Shear Encapsulation of Multiple Drugs. *Low G Journal, Vol. 9*, p. 1, March 1998
5. *Agência Espacial Brasileira (AEB), SBN - Quadra 02* - Bloco J Ed. Eng. Paulo Maurício, 5° andar Cep. 70040-905, Brasília, Brasil, http://www.agespacial.gov.br
6. Tomatos in Space, Historic Glenn Launch Seeds Student Space Experiments," *Spotlight Magazine, Lockheed Martin Technology Services, Volume 7*, Number 1, 1999
7. International Space University: ISU, Boulevard Gonthier d'Andernach, 67400 Illkirch-Graffenstaden, France, http://www.isunet.edu
8. La Tete Dans Les Etoiles, news article, *Dernières Nouvelles d'Alsace*, 1998
9. *First Brazilian Seminar on Space Education, 1957-1997: 40 Years of the Space Era*, Centro Técnico Aeroespacial, Instituto de Aeronàutica e Espaço, São José dos Campos, SP Brazil, October 21-22, 1997
10. Centro Técnico Aeroespacial, Instituto de Aeronàutica e Espaço, São José dos Campos, SP Brazil, http://www.iae.cta.br
11. http://www.lerc.nasa.gov/othergroups/pao/html/microgex.html
12. http://quest.arc.nasa.gov/smore/background/microgravity/mgintro.html
13. http://www.ITAspace.com
14. Watanade, Tokuji, Kishi, A.: *The Book of Soybeans: Nature's Miracle Protein*, 1982

Art Module "MICHELANGELO":
A Possibility for Non-Scientific ISS Utilization

C. Wilp, ART and SPACE, Friedrich-von-Spee-Str. 47, 40489 Düsseldorf, Germany

e-mail: artandspace@t-online.de

B. Bratke, DaimlerChrysler Aerospace AG, BEOS, Postfach 286156, 28361 Bremen, Germany

e-mail: burkhard.bratke@ri.dasa.de

Abstract
For the consumption as well as the production of art, the feeling of microgravity is expected to be an as yet unknown generator of human creativity. It is proposed that an art module, named "MICHELANGELO", be added to the International Space Station (ISS) after the technically designated elements are completely assembled and the operational activities have bgun. This module would be equipped with nothing but its life support systems; no racks or other equipment should disturb the available space within it. Its utilization would be completely devoted to art activities. For its realization private investors could choose between several different possibilities.

1. Introduction

According to current planning the assembly of the International Space Station ISS will be finished in the year 2004. It will then consist of a number of habitable modules, and will include three nodes, a service module, laboratory modules from four different countries and logistics modules. All the necessary elements that support its technical tasks and human work on it will be there. But how about the elements or provisions for recreational activities for the humans living on the ISS? Does the currently planned range of activities for a permanently inhabited human outpost in space reflect non-work life on Earth in an appropriate way? The answer can only be "no". Besides sporting activities the creation, performance and consumption of art represent a fundamental part of human life and humans living together anywhere on (or outside) the Earth. Therefore it is essential that the astronauts' time schedules leave time for art. This pastime could be spent nowhere better than in an art module, which could be added to the ISS. For the consumption as well as the production of art, the feeling of microgravity is, as yet, an unknown generator of human creativity.

2. The Idea

It is suggested that an art module, named "MICHELANGELO" (see Fig. 1), be added to the ISS after the technically designated elements are completely assembled and the operational activities have settled.

G. Haskell and M. Rycroft (eds.), International Space Station, 127-130.
© 2000 *Kluwer Academic Publishers.*

Figure 1. Collage of images for the art module "MICHELANGELO", with the first author in the foreground

This module would be equipped with nothing but its life support systems; there would be no racks or other equipment to disturb the available space within it. Its utilization would be completely devoted to art activities like painting, drawing, sculpting, producing of videos or movies (e.g. Mir – The Movie, filmed on the ISS, in case Mir is not available for that project anymore), designing fashion, writing literature (e.g., consider the Japanese journalist on Mir in 1996), creating music or performing dances, etc.. The microgravity of parabolic flights has been experienced by Charles Wilp to charge, subconsciously, batteries of creativity. The results of art activities in space may be expected to be unique art works of high value.

At first the astronauts themselves could use this location; however, it is also possible, though, that artists from Earth could work for a certain time on the ISS, thus becoming "ARTronauts".

Also entertainment, advertising and early space travel might be feasible non-scientific leisure activities in the art module. It is to be expected that the chance to have a module completely free from science and technology, to be used only for cultural purposes in weightlessness, would challenge the curiousity and sense for sensation of many people on Earth. The increased media attention during the recent Space Shuttle flight of Senator John Glenn can be seen as an indicator of this, even though most of that mission was devoted to science and research.

3. The Plan

The realization of this idea does not seem to be that far away, on its second view, as it seems to be on the first. A dedicated development for this module is not necessary. For its procurement private investors could choose between different possibilities. The Italian space company Alenia, for example, is currently building six structurally very similar flight modules for the ISS: the European laboratory module COLUMBUS, the Italian ISS contribution MPLM (3 flight units) and the ISS nodes 2 and 3. Boeing and Russian companies surely have similar elements available that could be relatively cheaply rebuilt to serve as an art module. ESA's ATV module might be another alternative for a certain period of time. The adaption of a no longer used Spacehab module for attachment to the ISS could be yet another possibility. After all, in the year 2003 the first Spacehab module will already celebrate its tenth anniversary in space. The utilization of inflatable structures, like the Transhab module currently being developed, opens up further options.

The transportation of the art module into space could be fairly inexpensive, if it can be used as a logistics carrier on its way up. Its adaption to the ISS and the resources used for it would then have to be bought from the ISS operator. Here, it is expected that these costs could be covered by renting the module to interested companies, consortia or even individuals, who are interested in creating or performing works of art in space, and including advertising activities. The selling of space art works should be a second source for financing.

4. Conclusion

Even though, at the moment, no sound and waterproof business plan can be presented for the idea outlined here, there exists great confidence in the concept and its commercial realization during the first decade of the new millennium. Contacts with relevant industrial partners have been established already.

Acknowledgements
The authors gratefully acknowledge the contributions of Martine Kerguel, ESA crew doctor, and Jean Pierre Haignere, Eurokosmonaut and pilot of the Caravelle airplane dubbed "Orbitic 22", who provided the first author with the direct feeling of microgravity for the production of art. Furthermore, Ingrid Schmidt-Winkeler is acknowledged for her invaluable support.

Space Education and Space-Based Education: The Russian Experience

O. Zhdanovich, International Space University, Strasbourg Central Campus, Parc d'Innovation, Boulevard Gonthier d'Andernach, 67400 Illkirch-Graffenstaden, France

D. Pieson, Moscow Aviation Institute Aerospace School., 4.,Volokolamskoe sch., GSP, Moscow 125871 Russia

e-mail: zhdanovich@isu.isunet.edu, pm@glasnet.ru

Abstract
        This paper discusses the experience of space and space-based educational programs developed in Russia from school to university level, including the Russian National Space Education Program, as well as public outreach programs carried on Russian TV/radio channels in the Soviet Union and continued in Russia during the last two decades.

1. Introduction

Russia already has more than 25 years of experience in long duration manned space flights, starting with the Salyut generation and followed by the Mir Space Station. Space education issues have been of concern within the professional community practically since the beginning of space programs. Significant efforts were made by the enthusiasts to introduce astronautics and space research issues into the regular education program as well as to utilize space technology features for educational purposes. However, there was not any public foundation that can unite all schools and universities working in aerospace education together. The NASA idea of the first Teacher in Space, with Sharon Christa McAuliffe, a member of the ill-fated Challenger crew, sped up the processes of creation of such an organization in the Soviet Union. At the end of 1988, the All-Union Youth Aerospace Society "Soyuz" was founded. It united many young professionals interested in space. As a result, many of those people who now develop the Russian National Space Education Program, publish the magazine Cosmonautics News ("Novosti kosmonavtiki") and produce TV/radio programs with the Videocosmos company, participated in the creation and development of that society.

2. Space Education in the Soviet Union before Mir Space Station

Space education has strong roots in the Russian national education system. The "Sputnik shock" of 1957 was caused to a large extent by the developed system of secondary and university level education in cosmonautics-related

131

G. Haskell and M. Rycroft (eds.), International Space Station, 131-135.

fields as well as an elaborate system for the motivation of talented youth to work in the aerospace arena.

As space exploration led to extensive studies aboard the Salyut Space Stations, the idea has been to attract young people to the design and planning of on-board experiments. In November 1981, a competition for school students (aged from 7 to 18 years old), Malyi Intercosmos ("Small Intercosmos"), was held with participants from the USSR and Eastern Block countries. The leading role in the organization of this competition was taken by the Moscow Palace of Pioneers (now Center for Youth Creativity) with Boris Pshenichner. The original idea was to attract students to develop in-space experiments. Later the contest as well as a number of similar (for example, annual Cosmos) competitions has been arranged for future spacecraft projects and a models' competition. As a result, these competitions have become a good student recruitment base for the leading aerospace universities.

The extension of this recruitment base was one of the reasons for the following event. As mentioned above, in 1988 the All-Union Youth Aerospace Society "Soyuz" was organized, chaired by cosmonaut Alexander Serebrov (now re-named the All-Russian Youth Aerospace Society). This society united schools and universities working in space education as well as established contacts with aerospace societies in the USA, Japan and China. The idea of that society is to find talented students interested in aerospace subjects at school level and to help them to enter aerospace universities. The Society has organized a variety of competitions mainly for school children in various topics of space related disciplines, including even a competition of space-related drawings. The winners have had the opportunity to visit youth aerospace camps in the USA and Japan.

3. Educational and Public Outreach Programs from the Mir Space Station

3.1 Lessons from Space

Cosmonaut Alexander Serebrov, President of the "Soyuz" Society, started space lessons for school children in 1990, during his flight on board the orbital Space Station Mir which he continued during his next mission in 1994. The first space lesson from Mir was dedicated to the memory of Sharon Christa McAuliffe, the first teacher in space, a member of the Challenger crew whose flight into space lasted only 76 seconds. Space lessons were dealing with the life of the cosmonauts on board Mir, crystal growth in conditions of zero-g and monitoring the Earth from space. The "Soyuz" Society organized a call for themes of space lessons in nation-wide specialized newspapers for school teachers. Teachers with the best proposals participated with their students in space lessons in the Mission Control Center.

3.2 TV Programs Based on Mir Videography

In 1990 the Videocosmos company was organized; it specializes in the covering of space exploration by developing TV/radio programs, producing video-documentary films for/about the Russian space industry and publishing a monthly magazine "Cosmonautics News" ("Novosti kosmonavtiki"). In 1992 it organized a joint programme on space-related topics with the famous Russian program for teenagers, "Maraphon-15". The program covered various aspects of the cosmonaut profession, showed the latest space news all over the world, as well as an astronomy TV serial produced by Canadian TV and translated into Russian. Cosmonaut Musa Manarov organized a geographical contest for young people in three programs of space for "Maraphon-15". This contest was based on video filming of the Earth from Mir during his one year mission in 1990-1991. In the first program Musa showed three easily recognised regions of the Earth. He suggested that children should work at home with maps and find the names of the places shown. In his second and third programs cosmonaut Manarov showed three cities and three lakes. The first winner of this TV geographical contest had the opportunity to visit the NASA Johnson Space Center in Houston, USA, and other winners received various presents.

Videocosmos also produced 10 to 15 minute video-films based on video recordings made aboard Mir, for example, as "Hotel halfway to the Moon" covering all the main issues of manned spaceflight to Mir, as well as "Flesh of the Earth" covering the Earth's environmental problems caused by human activities (such as air and coastal pollution) including the Gulf War as seen from space.

In November 1994 Videocosmos organized a TV bridge between Mir and French TV. Young people from France were discussing various aspects of life on board a Space Station with cosmonauts Alexander Serebrov and Vasily Tsibliev. In 1993 Videocosmos produced a documentary in 13 episodes for TV, "Red space", covering all the major milestones of the Soviet/Russian space program from the time of Kostantin Tsiolkovsky up to today, and produced a CD-ROM on "Soviets in space".

3.3 Joint Multidisciplinary Programs for Decisionmakers and Cosmonauts

In 1992 Professor Vitaly Gridin from the Moscow Academy of Oil and Gas developed an idea of joint space-ground studies and the management of natural resources of various regions of the Russian Federation from space. For traditional remote sensing techniques this approach is not new. The novelty of the idea of Prof. Vitaly Gridin is that the study and management of regional natural resources is done by joint multidisciplinary teams which include members of the regional administration, medical doctors, geologists, meteorologists, etc., including cosmonauts. The project starts during the cosmonauts' pre-flight training instructing them in the specific natural features and human activities of particular regions. With such projects it is possible to educate regional decision-makers as to how space technology can be useful for the particular region and well as educate cosmonauts more deeply in environmental issues. In 1993-1994, such a project was carried out for the Orenburg region. Three cosmonauts obtained a second Master's degree, in Environmental Science, from the program developed between the Cosmonaut Training Center and the Moscow University of Geodesy and Cartography.

4. Russian National Space Educational Program

In spite of a number of separate efforts, until recently there was no country-wide program for space education, neither in the Former Soviet Union (FSU) nor in the recent Russia. To fill the gap, in 1996 a study was initiated by the Russian Space Agency (RKA) to develop and implement the Russian National Space Education Program [Reference 1]; the Moscow Aviation Institute's International Center for Advanced Studies, COSMOS, was assigned by the RKA as the prime contractor to develop this program.

According to the Program's definition, space education is a wide spectrum of activities:

- to recognize the human role and place in space exploration, as well as the relationships between Earth and space phenomena
- to understand the meaning and role of space research and space applications in human life
- to use the achievements of cosmonautics in the different fields of science and economics
- to master the space-related professions
- to facilitate decision making in fields relating to space activities.

The major *directions* of the space education activities covered by the Program are:

- Development of the space education system for the general public, including public relations (PR) activities. Development of the space education information and knowledge tools

- Deepening and expansion of the space-oriented courses within the pre-university education system. Professional development of primary and high school teachers in the space field
- Development of methods and managerial structures for the universities which provide both the aerospace industry with human resources and the country with new intellectuals
- Improvement of the professional development of, and postgraduate curricula for, the aerospace industry and science.

The six *subject fields* covered as a set of subprograms are:

- Space education of the population and public relations
- Primary space education
- High school and university space education
- Postgraduate and professional development education in rocket and space technology
- Distant learning space education
- International co-operation in space education.

Within the National Space Education Program framework, Space Stations and satellite-based educational programs are of special importance. Also, the public outreach and students' professional orientation aspects aboard the ISS are important parts of this Program.

5. Conclusions

The Russian experience in the development of space and space-based education programs as well public outreach programs aboard Mir will be of great help for the development of such programs aboard the International Space Station.

Acknowledgements
I would like to express my gratitude to Vladimir Semenov, President of Videocosmos Inc., Russia, for providing videomaterials of TV programs based on Mir videography for the oral presentation of this paper.

References
1. Alifanov, O., Senkevich, V., Usyukin, V., Khokhulin, V., Doroshin, V, and others: Russian National Space Education Program, a report for the Russian Space Agency, Moscow, 1998

Report on Panel Discussion 3:

Education and Public Awareness

L. Higgs, C. P. Karunaharan, International Space University, Strasbourg Central Campus, Parc d'Innovation, Boulevard Gonthier d'Andernach, 67400 Illkirch-Graffenstaden, France

e-mail: higgs@mss.isunet.edu, karunaharan@mss.isunet.edu

Panel Chair: R. Oosterlinck, ESA Headquarters, Paris, France

Panel Members:

J. D. Burke, Senior Member, Technical Staff JPL (Retired on-call), USA
V. Cassanto, ITA Inc., USA
N. Ochanda, University of Nairobi, Kenya
M. Uhran, NASA, USA
C. Wilp, ART and SPACE, Germany
O. Zhdanovich, International Space University

Opening the discussion, the chairman stated that, in ESA, it is mandatory to give, not only postdoctoral fellowships, but also education for young people (science and technology to primary and secondary schools). There is now an inter-agency working group for education.

Walt Disney has shown an interest in space and may provide money and education. What suggestions would the panel provide? **V. Cassanto** opened the replies by stating that imagery and animation could be integrated in films made in space and broadcast from the ISS to Earth. **O. Zhdanovich** proposed showing how science is done in space. **J. Burke** said that Disney could make toys for education, similar to Sojourner, the Mars surface rover. **R. Oosterlinck** made the point that young people's dreams and visions were important and that showing images helps to create dreams.

In the 1950's Disney had visualised Von Braun's ideas. In education, what is the vision of the future? **J. Burke** remarked that choreography for the ISS, with its low gravity, and dancing gives the young ideas of what they could do in space. **R. Oosterlinck** said that the young want to solve problems and the ISS could motivate them to solve the mysteries of space; **V. Cassanto** agreed. **C. Wilp** gave an example of a seven year old girl involved in a parabolic flight who wanted to be a cook — her imagination was captured, and she wanted to cook for astronauts!

G. Haskell and M. Rycroft (eds.), International Space Station, 137-138.
© 2000 *Kluwer Academic Publishers.*

What is the co-operation between space agencies as ISS partners, and what is ISU's role in education? **R. Oosterlinck** portrayed a public relations role, which would be a forum for the meeting and exchanging of information. He noted that in Europe it was difficult to introduce changes to the curriculum, but ESA is taking steps to introduce educational "tools"; however, Japan is very advanced in this field. **J. Burke** added that NASA JSC in Houston has a similar programme, and that the Web site has education and activity spin-offs from the Space Shuttle programme. He recommended that the ISU could have a student Design Project or Team Project to this effect and place itself as a "broker" connecting educational groups. ISU was the right forum because of its interdisciplinary, intercultural and international approach. **O. Zhdanovich** suggested that ISU, its affiliates and alumni should be involved in research for the interest of the general public. Great Britain is at the forefront of space education through its "Space Days", boasted **C. Wilp**.

The next question was directed at **N. Ochanda**. How would the above scenario apply in Africa and would the same technology apply? He answered that a link should be created at a basic level between the universities and the general public. Concluding this inspirational session, the question was raised of why Africa was not taking part in the ISS —was it due to the fact that although they had the technology they did not have the users? In reply, **N. Ochanda** pointed out the need to increase public awareness. **R. Oosterlinck** reminded the audience that space spin-offs were already being implemented in the fields of telecommunications and education, as well as using solar cells domestically in tropical countries.

In his closing remarks, **R. Oosterlinck** rejoiced at the fact that the panel was from a variety of different generations and cultures.

Session 4

Innovative Approaches to Legal and Regulatory Issues

Session Chair:

A. Farand, ESA

Legal Environment for Exploitation of the International Space Station (ISS)

A. Farand, European Space Agency, 8-10 rue Mario-Nikis, 75736 Paris, France

e-mail: afarand@hq.esa.fr

Abstract

This presentation focuses on the legal environment created by the Intergovernmental Agreement (IGA) and the four related Memoranda of Understanding (MOUs) for carrying out International Space Station (ISS) operation and utilisation activities, over the useful lifetime of the flight elements provided by the ISS Partners. It addresses the issues to be dealt with in the arrangements which will be concluded for enabling the various categories of users, from whatever field of activity in either the public or private sector, to exercise utilisation rights belonging in the first place to the Partner's Cooperating Agencies, while attracting funding from sources other than States' contributions. The presentation also reviews the efforts being made by the European Space Agency (ESA), on behalf of the European Partner, to prepare for efficient and effective use of the European utilisation allocation of the ISS, despite the difficulty of reconciling competing interests at both national and Agency levels. Finally, the presentation examines different aspects and implications of the new ISS exploitation programme subscribed in May 1999 by the ESA Member States concerned, the most ambitious and complex exploitation programme ever undertaken by the Agency, with the added constraint of avoiding exchange of funds between ISS Partners.

1. Introduction

The general framework put in place for cooperation on the International Space Station project comprises three layers of agreements, which are explained in detail: a multilateral intergovernmental agreement (IGA); four Memoranda of Understanding between the designated Cooperating Space Agencies; and the implementing arrangements already concluded or to be concluded over the duration of the cooperation. The general rule is that a State can exercise its control and jurisdiction only on its territory and in its air space; the IGA therefore constitutes the basis on which the signatory States are allowed to extend their national jurisdictions and controls into a facility located in outer space. Before its ratification of the IGA, a State will make sure, through adoption of appropriate legislation for example, that its national legal system is compatible with the commitments which it has subscribed in the IGA and take appropriate means to ensure that its national law can apply over the flight elements and personnel which it provides to the project. Furthermore, in elaborating a comprehensive legal regime governing activities taking place on board the International Space Station, the States concerned have not created a new body of laws applying to the ISS; they have rather made links between the ISS, or more precisely its modules and personnel, and their territories so as to authorise the application of their national laws to a given situation.

G. Haskell and M. Rycroft (eds.), International Space Station, 141-153.

2. The Three-layer Legal Framework

2.1 The Intergovernmental Agreement

On 29 January 1998, the representatives of fifteen States, i.e. the United States, Russia, Japan, Canada and eleven Member States of ESA, signed in Washington an intergovernmental agreement (referred to as the IGA) concerning cooperation on the civil International Space Station. This agreement which, when brought into force, will replace the 1988 IGA, not only formalises Russia's integration in the partnership but also confirms major changes in the Partners' contributions and a dramatic evolution of the rules put in place for this cooperation.

Because of the expected 15-year duration of this project, the corresponding multi-billion dollar envelope to be spent by each Partner and the fact that carrying out a project of such magnitude requires arrangements in some fields of jurisdiction that are clearly beyond those falling under the responsibility of Space Agencies, it was decided not to limit the legal instruments to Agency-level MOUs as for previous cooperative endeavours but rather to involve the Governments wishing to participate in the project through the conclusion of an international agreement. The IGA sets out the general principles for carrying out this cooperation, including those governing the parties' conduct in outer space. It establishes "a long-term international cooperative framework among the Partners, on the basis of genuine partnership, for the detailed design, development, operation, and utilization of a permanently inhabited civil International Space Station for peaceful purposes, in accordance with international law". The IGA makes a distinction between Partner States and Partners which is quite innovative in terms of international law; this is realised when one looks at particular responsibilities reserved for Partners and others for Partner States in the IGA. This is a distinction of particular importance for Europe. There are fifteen Partner States but only four Partners in the project because the eleven European States are grouped, for the purpose of conducting this cooperation, under the umbrella designation of the "European Partner".

The signature ceremony on 29 January 1998, which followed more than four years of rigorous bilateral and multilateral negotiations, can be characterised as a major milestone in the international partnership. In addition to a fairly broad legal regime developed in the IGA itself for the conduct of Space Station cooperation, very innovative rules have been drafted to govern such things as the development and utilisation of the Space Station, and the management and financing of the Partners' programmes and the international programme made up of the Partners' combined contributions. Although the original concept of an integrated Space Station has been preserved in this negotiation process, many features of the cooperation have been modified, generally for the sake of underlining the genuine partnership concept, or have evolved considerably from what was envisaged at the outset.

During the negotiations held between 1994 and 1997, the original Partners could justify their acceptance of a number of IGA provisions confirming the lead role of the United States in the international programme not only because of the overwhelming importance of its contribution to the programme but mainly because of the need to provide for a clear line of command and control in this endeavour. Throughout the negotiation process, on the strength of its long experience of long-duration human spaceflight, Russia pressed for recognition in the Space Station Agreements of a role which would reflect the qualitative and quantitative importance of its contributions to the International Space Station programme. In the renegotiation of the IGA, this overarching Russian requirement was a factor as important as the U.S. leadership had been during the original IGA negotiations of 1986-1988 in establishing a particular balance between the Partners, this being accomplished without prejudice to the genuine partnership concept. As a result of the more recent negotiations, the lead role of the United States, and almost all its original responsibilities for overall programme management and coordination, have been confirmed in the IGA. However, a large number of changes were made to reflect the new technical reality brought about primarily by Russia's contributions but also by Europe's redesign of its original contributions to the project and its insistence on recognition of specific activities, including the periodical correction of the Station's orbit using the ESA-developed Automated Transfer Vehicle in conjunction with Ariane 5.

2.2 The Memoranda of Understanding

Also on 29 January 1998, the Head of NASA and of the Heads of the Russian Space Agency, the European Space Agency and the Canadian Space Agency respectively signed a Memorandum of Understanding (MOU) containing detailed provisions for implementation of Space Station cooperation. While it was originally envisaged that a fourth similarly-worded MOU with the Japanese cooperating Agency would be signed only after ratification of the new IGA by the Japanese Diet, NASA and the Government of Japan, representing a series of Japanese agencies charged with different aspects of the cooperation, changed their approach and signed their MOU on 24 February 1998, thus enabling the Diet to examine this MOU together with the IGA.

When brought into force, i.e. after the Parties notify each other that their internal procedures required for this purpose have been completed, the MOUs signed on 29 January and 24 February 1998 will replace the three original ones signed in 1988. The four new MOUs concern the detailed design, development and operation of a manned civil Space Station. A memorandum of understanding is generally not considered to be an agreement generating rights and obligations at international law for its signatories, although this does not exclude the possibility of remedies provided for under a Partner State's legal system being applicable on the basis of a memorandum of understanding if, for example, a party to it failed to discharge its obligations appropriately. The memorandum of understanding is considered to be a type of arrangement that registers a political and moral commitment on the part of an international organisation, a government, or a constituent part of the latter, to conduct itself in a certain way. Because of their close links with the IGA, it would appear that the Space Station MOUs will have acquired the status of international agreement, as an exception to the general practice in this field.

It is interesting to note that the multilateral bodies established for the management of Space Station, such as the Multilateral Coordination Board (MCB), the top-level body in charge of coordinating the activities of all Cooperating Agencies related to the operation and utilisation of the Space Station, are provided for in the MOUs, which are bilateral instruments between NASA and each of the other Cooperating Agencies. The pattern of cooperation put in place through the MOUs has been referred to as the "hub and spoke" approach, similar to the pattern adopted for air transport in a number of countries.   In this particular instance, NASA is the hub and all the other cooperating agencies are "spokes": one consequence of this pattern is that a commitment made in a given MOU by a Cooperating Agency in favour of others has to be reflected in the other relevant MOUs, with this commitment "transiting", so to speak, through NASA, which is a party to all the MOUs. It is also the MOUs that define the five-year "sliding" planning process for the operation and utilisation of the ISS, including the procedure put in place by the Cooperating Agencies for approving the corresponding documentation

2.3 The Implementing Arrangements, and Other Arrangements

The third layer of international instruments is represented by the "implementing arrangements" referred to in Article 4 of the IGA.   These arrangements, relating to implementation of the Parties' obligations or the exercise of their rights, as spelled out in the MOUs, are subject to the MOUs and thus NASA always has to be a party to them. The IGA and the four recently signed MOUs contain numerous provisions calling for the conclusion of implementing arrangements and, in that sense, the IGA and MOUs are only the tip of the iceberg of legal instruments that need to be put in place by the Partner States and the Cooperating Agencies. At this stage, only one implementing arrangement has been concluded between ESA and NASA: it relates to the barter between the NASA launch of the ESA-developed Columbus Orbital Facility (COF), using the US Space Shuttle, and the development by ESA, and delivery to NASA, of the ISS Nodes 2 and 3 and equipment to be used on board the US laboratory.

Article 4.2 of the IGA which establishes this hierarchy between Space Station Agreements (IGA, MOUs and implementing arrangements) is silent on the other arrangements and agreements that may be concluded between Partners for the purpose of furthering Space Station cooperation. One example of these is the Memorandum of Understanding concluded in 1990 (and amended in 1997) between NASA and the Italian Space Agency (ASI) for the development by ASI of the mini pressurised logistics module (MPLM) which, under the MOU, is a NASA-provided element of the Space Station. Another example of an arrangement not provided for in Article 4.2 of the 1988 IGA, which is arises from Russia's arrival in the partnership is the arrangement signed in 1996 between ESA and the Russian Space Agency (RSA) for the delivery by ESA to the RSA of a European external robotic arm (ERA) to be used on the Russian segment of the Space Station.

3. The Partner States' Obligations

For the purposes of discharging almost all its ISS-related responsibilities and exercising its rights, the European Partner is acting through ESA.[1] However, a number of responsibilities, for example in the fields of customs and immigration and of criminal jurisdiction, are exercised directly and exclusively by the States themselves.

The explicit obligations contained in the IGA which have been assumed by the European Partner, and consequently by the European Partner States jointly and severally, and which have a bearing on utilisation of the Space Station, are: (a) to provide the elements listed in Section 2 of the IGA Annex (flight elements, ground elements and other elements); (b) to take responsibility for the operation of the elements which it provides and to develop and implement procedures for operating the Space Station in a manner that is safe, efficient, and effective for

---

[1] A number of IGA provisions confirm that the European Partner's obligations, and therefore the European Partner States' obligations, will be discharged (and the corresponding right will be exercised) through ESA:(I) the paragraph in the preamble referring to the ISS programme, (ii) Art. 4.1 which designates ESA as the European Partner's Cooperating Agency, and Art.4.2 which provides that "the Cooperating Agencies shall implement Space Station cooperation...", (iii) Art. 6.2 which provides that the European Partner entrusts ESA with ownership of the elements and equipment developed and funded under an ESA programme as a contribution to the Space Station, its operation or utilisation, and (iv) Section 2 of the IGA Annex which lists the elements (flight, ground and other elements) to be provided by the European Partner, through ESA.

Space Station users and operators; also, to sustain the functional performance of the elements which it provides; (c) to make available launch and return transport services for the Space Station; and (d) to bear the costs of fulfilling its responsibilities under the IGA.

4. Utilisation of the Space Station

The basic principles for utilisation of the Space Station are laid down in Article 9.1 of the IGA:

> "Utilization rights are derived from Partner provision of user elements, infrastructure elements, or both. Any Partner that provides Space Station user elements shall retain use of those elements, except as otherwise provided for in this paragraph. Partners which provide resources to operate and use the Space Station, which are derived from their Space Station infrastructure elements, shall receive in exchange a fixed share of the use of certain user elements."

The share of the use of user accommodations, such as pressurised laboratories, to be retained by the Partner providing that accommodation is expressed in fixed percentages in the MOUs. To be more precise, ESA will keep 51% of the user accommodation on the European pressurised laboratory and Japan's Cooperating Agency will retain the use of 51% of the user accommodations on the Japanese Experiment Module (JEM). The remaining 49% of user accommodation in the COF and JEM are attributed to those Partners providing infrastructure resources to ESA and Japan's Cooperating Agency (referred to in the MOUs as "the GOJ"), essentially NASA but also the CSA, which is providing the Remote Manipulator System (RMS) as an infrastructure element.

A second step in the understanding of the principles applicable to utilisation of the Space Station is an examination of the approach taken in the allocation of Space Station resources. First, an agreement has been reached between the original Partners and Russia based on the premise that Russia on the one hand and the other Partners on the other retain utilisation of their own contributions to the Station and seek to offset only those items that cross the interface. This of course has many implications with regard to the sharing of ISS resources and the treatment of common operations costs involving exchanges between the Russian segment and the segment of the Space Station composed of elements (the Alpha segment) provided by the other four Partners. The Partners have nevertheless laid strong emphasis on the need for the closest possible adherence to the philosophy of an integrated International Space Station and the rules underpinning that philosophy in the Space Station Agreements.

By way of illustration, it was decided that, for the purposes of sharing utilisation, the Russian Partner would keep 100% of utilisation of its own modules, thereby recognising that the infrastructure element supplied to the Station by Russia for its own benefit and that of the other Partners would enable it to accumulate up to 100% of the utilisation rights in its own modules. This calculation has the advantage of avoiding a debate on the relative value of the utilisation and infrastructure elements supplied by Russia as a proportion of the Space Station as a whole. This means that the percentage agreed, on the basis of 100% within the Alpha segment, among the founding Partners could be retained for the purpose of sharing available resources. The MOUs specify the precise percentage of resources to be allocated to each Cooperating Agency: for example, ESA's share has been fixed at 8.3% of the resources available for sharing on board the Alpha segment.

Establishing a direct link between the allocation of resources and the financial responsibilities of the Cooperating Agencies, Article 9.3(a) of the ESA/NASA MOU provides that:

> "NASA, ESA and the other partners will equitably share responsibilities for the common system operations costs or activities, that is the costs or activities attributed to the operation of the Space Station as a whole.... RSA will be responsible for the share of the common system operations costs or activities corresponding to the operation of the elements it provides. NASA, ESA, the GOJ and CSA collectively will be responsible for the share of common system operations costs or activities corresponding to the support of the operation of elements they collectively provide using the following approach: each will be responsible for a percentage of common system operations costs or activities equal to the percentage of Space Station utilization resources allocated to it ..."

In addition to the above-mentioned common system operations costs responsibilities, each Partner will be financially responsible for costs or activities attributed to operating and sustaining the functional performance of the flight and ground elements which it provides and the use of its user accommodations. To give an idea of the scale of costs to be borne by each Partner, it can be reported that ESA has estimated that the total exploitation costs over a period of 10 to 11 years would be of the same order of magnitude as the total development costs of its contributions, two thirds of that sum being devoted to discharging common system operations responsibilities. This explains the efforts put by the European Partner into persuading its Partners of the need to lay down transparent financial rules for the cooperation.

5. Financial Rules Applicable to the Cooperation

A significant interest of the European Partner in the IGA and MOU negotiation process which extended from mid-1994 until end of 1997 were of a financial order and resulted from directives and guidelines given by the participants at the ESA Council meeting at ministerial level held in Granada in November 1992. The European Partner therefore proposed to amend Article 15 of the IGA on Funding with a view to formalising two concepts: (a) the offset concept, according to which a Partner would be able to meet its share of the Station's common system operations costs by supplying goods and services produced by itself, and (b) the concept of the "not-to exceed figure", which would involve the establishment of procedures administered by the management bodies for containing the common system operations costs within predetermined and agreed levels, thus imposing a ceiling on these costs. This would enable a Partner to know the full extent of its commitment sufficiently in advance to plan its expenditure accordingly.

These two concepts are linked to agreement among all the Partners on the setting-up of a fleet of spacecraft supplied by four of the five Partners to meet all the Station's transport requirements. This change, which was unavoidable with Russia's arrival in the partnership, represents a significant departure from the situation outlined in the IGA signed in 1988 where the U.S. Space Shuttle was the only space transport system to be used for the cooperation. With Ariane-5, operating in conjunction with the ATV, the European Partner is in a position to discharge its share of common costs in a worthwhile manner, given that space transport is going to account for some 80% of the Station's common operations costs. Much of the discussion between the European negotiators and their counterparts centred on the type of assurance which the European Partner could be given at this stage by the United States and the other Partners to the effect that Ariane-5/ATV, deployed on Station orbit reboost missions for instance, and other European services would indeed be used to offset the whole European share of common system operations responsibilities, so that cooperation could be established on the basis of "no exchange of funds" between the European Partner and its Partners.

6. How the European Partner is Organising its Utilisation of the ISS

6.1 The ESA ISS Exploitation Programme

At the ministerial meeting of the ESA Council in Toulouse in October 1995, the ten ESA Member States participating in the development of the European contributions to the ISS already decided on the rules to be applied in an appropriate programmatic framework for the subsequent operation and utilisation of the Space Station by the European Partner. Their motive for doing so at that early stage was that, at the time that a decision with significant financial consequences was being taken on the last step of the development of Europe's contributions to the ISS project, there was a need to make sure that there was a genuine political and financial commitment to operate and use this in-orbit infrastructure.

On 12 May 1999, also at a ministerial meeting of the ESA Council, this time held in Brussels, the same States finalised and brought into force the legal instrument setting up an ESA ISS exploitation programme, confirming that their participation in that new programme would extend until the end of the useful lifetime of the European flight elements contributed to the ISS, envisaged for 2013. This was accompanied by the participating State's subscription of the financial sub-envelope covering the two-year first step of early activities and their confirmation of a complementary provisional financial commitment for the following three-year period. The above financial sub-envelope, and therefore the participating States' subscriptions, have been broken down between fixed and variable costs, the former being more closely related to the operation responsibilities of the European Partner, including its share of common operations costs, and the latter being related to its utilisation interests in the ISS. However, this distinction between fixed and variable costs will remain rather artificial until actual utilisation of the Station by the users of the European Partner States, from 2004 onwards.

Although it is reassuring that the start of this new exploitation programme should ensure that the corresponding activities will be carried out on a stable financial basis over the long term, it should be stressed that this recent decision was accompanied by very explicit directives given to ESA not only to make all reasonable efforts to bring this exploitation programme within more palatable financial parameters but also to find new funding sources.[2] This could be explained by the rather consistent past practice of the European Space Agency

---

[2] In this connection, the last paragraph of Resolution No. 2 adopted by ministers assembled at the ESA Council meeting at Brussels on 12 May 1999 reads as follows: "INVITES the Director General to identify and propose the best conditions and structures for promoting efficient and effective operation and utilisation of the various infrastructure elements such as the International Space Station and launchers developed by Agency programmes, and in particular to examine the scope for industrialising exploitation of the ISS, and submit a corresponding proposal to Council by March 2000 and STRESSES the need to execute operations and utilisation activities in partnership with other European entities, such as the European Commission, or with industry, commercial users and commercial operators and to involve the various user programmes of the Agency in those activities.

of transferring technology and facilities developed under its programmes to other international organisations or private entities for the purposes of their exploitation, so as to prevent a sizeable part of ESA's budget being diverted from its core mission of research and development in the space domain. This allows ESA to largely keep out of the space-operations business and concentrate on fostering new technologies and exploration, and to ensure that exploitation is carried out in the best possible environment, commercial or otherwise. The concern voiced by Member States' representatives at the Brussels meeting is also understandable when it is realised that the ESA ISS exploitation programme could represent a steady-state average of approximately 11% of ESA's total annual budget, with close to two-thirds of that sum being devoted to the operations side of exploitation. The Agency would then be left with no resources with which to enter into new fields of manned space activity.

Undoubtedly, the wealth of literature produced by the numerous forums that have considered various aspects of commercialisation of the ISS will be a good point of departure for ESA's own deliberation on the subject. Two aspects will be examined: (a) how to succeed in the transfer to the private sector of the largest possible portion of the operations side of ISS exploitation, from mission planning and maintenance to transport, communications and training, on the understanding that the Partner States and ESA retain their overall responsibilities under the ISS agreements, and (b) how to use a significant portion of ESA's ISS utilisation rights for development of new products and services by the private sector ranging, for example, from pharmaceutical companies and telecommunications firms to the entertainment industry and advertising agencies, through an appropriate marketing effort. Privatisation of operations could be justified only if, as expected, private companies are able to provide comparable levels of service and technical sophistication at a much lower cost. Before the Cooperating Agencies could envisage recouping some of the public's investment in the development of the ISS, their objective will be to recoup the costs directly attributable to use of the ISS facilities, then those related to maintenance of the infrastructure, referred to as "fixed cost" in the ESA programme.[3]

---

[3] In an article of 15 December 1998 published on the Internet by ABC News and entitled "ISS, Brought to You By", Peter N. Spotts lists the following obstacles to be cleared before commercialisation could be successful: (a) prohibitive launch costs, (b) too few flights and unacceptable lead time between applying for a flight and launch, (c) the lack of a basic price list for Space Station activities, as well as a central clearing house for proposals from the private sector.

6.2   *Different Aspects of European Utilisation of the ISS*

A part of the development programme decided upon in October 1995 and currently being implemented is devoted to preparations for European utilisation of the Space Station.   In this connection, ESA has already issued "Announcements of Opportunity" (AOs) addressed to potential European ISS users in different fields of research and development for the purpose of collecting proposals for experiments to take place on the ISS.   Since utilisation rights will not formally start accruing to ESA until the COF is verified in orbit, an event currently expected to take place at the beginning of 2003, in January 1997 ESA concluded with NASA a barter arrangement that will provide ESA users with utilisation opportunities as early as 2001 and 2002, on board the US ISS laboratory, in exchange for the provision by ESA of specialised laboratory equipment.

It has already been decided that European utilisation of the ISS will be organised in consecutive phases, so as to allow for necessary adjustments due to changes in the programme, the need to inject some flexibility into the actual implementation, experience gained and promotional aspects.   The first utilisation phase will be based on the early utilisation opportunities secured from NASA as just mentioned.   An annual European utilisation plan will be submitted for approval to the Manned Space Programme Board, the ESA delegate body in charge of manned programmes.   This plan, covering the totality of the share of ISS utilisation accruing to the European Partner, will be prepared by a recently created working and advisory body called the European Utilisation Board (the EUB); this preparation will include a review of proposals by experts in the field and prioritisation of proposals according to a number of recognised criteria.

More detailed utilisation rules will be developed in the framework of ESA over the coming years so as to regulate the specific conditions for user access, including access by space programmes carried out by individual ESA Member States, and third-party users, which comprise industrial and commercial users. These rules will also address such matters as: (a) the allocation of resources to different disciplines among the various user categories, (b) the procedure for issuing Announcements of Opportunities, (c) the criteria to be applied in the selection of facilities, experiments and investigations aboard the ISS, (d) the means of protecting the intellectual property rights of the different categories of users, (e) the information and data access policy, and (f) the charging policy to be applied to users originating from ESA Member States other than those participating in the programme and to third parties.   Finally, ESA will need to ensure the coordination of the European utilisation plan with those of the other Cooperating Agencies, consistent with applicable MOU procedures.

6.3   Arrangements Between the Users and the Cooperating Agencies

As we have seen, the IGA and MOUs have established the broad framework for the basic rules which Partner States have agreed to apply to all Space Station-related activities, including utilisation activities. However, one could say that this legal framework is fairly general and permissive and, except for two elements mentioned below, does not impose any specific condition to the Cooperating Agencies in their dealings with potential users, in particular in relation to the protection of these users' intellectual property rights. In dealing with potential users of Space Station facilities, the interested Cooperating Agency shall include in the corresponding contractual arrangements (a) the cross-waiver of liability outlined in Article 16 of the IGA so as to ensure, on the basis of reciprocity, that the user is protected against claims that could be presented by the other Space Station Partners or their respective related entities,[4] and (b) an explicit obligation for the user to provide the Cooperating Agency with sufficient data on the payload or experiment it intends to provide so that the Agency will be able to discharge its own obligations towards the partnership, including those related to safety, with regard to that payload or experiment.

As far as protection of Space Station users' intellectual property rights is concerned, it should be mentioned that Article 19 of the IGA contains rules and procedures, related to markings, to protect against any retransfer to third parties of any data or goods which a Cooperating Agency is obligated to provide to another Agency if such data or goods are protected for proprietary or export control purposes. Also, Article 21 of the IGA, on Intellectual Property, establishes that patent laws of the Partner State having provided the flight element in which an invention has taken place shall apply to the patenting of that invention and also establishes a number of assumptions in case of claims, before a European jurisdiction, for patent infringement; these rules are very general, and essentially procedural, and do not affect the nature of the arrangements that can be made between a Cooperating Agency and its sponsored user for the sharing of benefits accruing from an invention resulting from Space Station work. It can be assumed that in most instances the invention resulting from the experiment will be identified or made from raw data after the experiment or payload is returned to Earth. Therefore, the invention will be patented in accordance with the rules applicable in the corresponding State.

---

[4] This, however, shall not prejudge the liability regime to be negotiated between the user and the sponsoring Cooperating Agency, although there also a cross-waiver of liability seems to be the relevant approach.

Finally, it could be recalled that the "black box" approach, and the need for the Partners not to be too intrusive on payloads provided by other Partners' users or by their own users if that has been specified in the sponsoring Partners' contractual commitment, was present in the mind of the 1988 IGA negotiators when they developed Article 20 of the IGA on Data and Goods in transit.

7. Conclusions

The rules established for Space Station cooperation will contribute to a certain emancipation — from a legal standpoint and compared with the current situation in which only the United States and Russia have the technical means and expertise to send a human being into outer space — of the Cooperating Agencies of Europe, Japan and Canada in their manned space activities. This emancipation will have a beneficial impact on all aspects of these activities. The development of Space Station rules will be a challenging task for the European Partner States, in particular because it calls for an effort of harmonisation between their national laws and regulations applicable to one aspect or another of Space Station cooperation.

As with almost all aspects of Space Station cooperation, more work lies ahead for the Cooperating Agencies after the signature of the new Space Station Agreements in January 1998. In particular, the Code of Conduct for the astronauts provided for in Article 11 of the IGA, which is not technically an implementing arrangement but could be seen as having a legal status somewhat similar to that of the IGA and the MOUs, is the most urgently needed document, and also a very complex one, yet to be developed by the Partners. In addition to the forthcoming negotiations between the Cooperating Agencies on various legal instruments called for in the IGA and the MOUs, these Agencies will have to develop individually not only their detailed utilisation rules but also their own approach, and the corresponding policies and rules, for proceeding with what has been referred to as the industrialisation of the exploitation of the ISS. This expression applies to a combination of privatisation of the bulk of ISS operations and commercialisation of a significant part of the available utilisation capacity offered by the Space Station.

Promotion of Industrial ISS Utilization by the German Space Agency

F. Claasen, P. Weber, H. Ripken, V. Sobick, German Aerospace Center, 53227 Bonn-Oberkassel, Germany

e-mail: Friedhelm.Claasen@dlr.de

Abstract
The International Space Station is a global cooperative programme between the United States, Russia, Canada, Japan and Europe for the joint development, operation and utilization of a permanently inhabited space station in low Earth orbit. Germany contributes 41% of the overall development costs for the European part of the ISS. During recent decades Germany has been strongly involved in manned space flight programmes and has gained extensive experience in microgravity sciences, engineering and management. Therefore Germany is aiming at a leading role in ISS utilization within Europe and ESA.
The ISS is a new orbital structure with the possibilities of longterm experimentation, servicing of hardware, short term and regular access, availibility of more energy and data transmission, realtime video and robotics. These new boundary conditions are very promising prerequisites, not only for the well established space community, but also for industrial and commercial activities in space by non-space industries. The ISS will become a tool like any other large scale laboratory facility on Earth. To make this new research facility known among non-space industry researchers and to get them "onboard" the DLR has initiated the project "Promotion of industrial users of the ISS". Within this, the relevant information is distributed via adequate media, e.g. workshops, periodical Newsletter, Internet, and potential users are addressed via national associations of sciences and of industrial companies in a framework of events (symposia, meetings, workshops). In the early utilization phase, ESA and the national agencies intend to provide for the costs of the flight, logistics and the required system operations. In principle the user has to provide his experiment hardware and to cover the expenses for the related ground-based research. Industrial requirements for ISS use are identified and have to be implemented according to the jointly to be agreed access rules of the ISS partners. Administrative and legal questions, e.g. proprietary rights, confidentiality, charging policies, advertisement rules, costs and schedules, have to be settled by clear and transparent international agreements. Within such a framework the ISS can be a valuable tool for profit-oriented industrial and commercial ventures.

1. Introduction

The International Space Station will be built by the five international partner organizations NASA, RKA, NASDA, CSA and ESA. The ten participating European nations are represented by ESA. Amongst them Germany contributes the major financial part, 41% of the overall development cost for the European part of the ISS. Future contributions for the exploitation phase, namely for operation and utilization of the ISS are not yet fixed, but ongoing European negotiations let us assume similar efforts from Germany [Reference 1].

G. Haskell and M. Rycroft (eds.), International Space Station, 155-162.

In the past Germany has gained broad experience in manned space flight projects: foremost with Spacelab, FSLP, D1 and D2, then with participation in many US shuttle missions (IML, MSL), with several cooperations with Russia using the Mir space station, and furthermore with cooperative ESA projects. Major experience has been achieved in microgravity sciences, engineering and project management. This experience can now be used to maintain a leading role in ISS utilization in Europe.

In April 1999, the German Chancellor Schröder chaired the opening ceremony of the Spacelab exhibition in Bremen. He demanded an active role from Germany to apply the knowledge gained in many years as an important international space partner, especially for the upcoming industrial utilization.

2. One Third of the ISS Resources for Industrial Utilization?

The first two elements of the ISS have been launched successfully last year and the construction is now an ongoing and continuous process. The station will be available for research very soon and that will open a completely new era of space investigations. A permanent crew can be expected shortly after the Service Module has been launched.

The German space industry has gained a broad know-how during the past 25 years. Now, shortly before the International Space Station is permanently operated, we have to provide well suited boundary conditions, especially for the non-space industries, enabling them to make the best use of this new research facility in space.

2.1 European Utilization Conditions

Within Europe the ESA utilization concept for the ISS is based on three utilization branches: fundamental and applied research, industrial/commercial users, and third parties. Announcements of Opportunity are open worldwide, and different selection processes apply for the different branches. These include established peer reviews looking for the best science in the case of fundamental research, evaluating market potentials for pure commercial experiments, and selection on a case by case decision for third parties. The recent definitions of user categories and ISS user access guidelines are to be found in the relevant documentation of the ESA Program Board Manned Space [Reference 2]. There is no *a priori* allocation defined for the different utilization branches, and the Manned Space Program Board will strive for a balance amongst the different disciplines.

One of the most difficult problems arises with proprietary rights for the resulting research data. According to existing overall ESA rules it is required that, for payloads flown on an ESA mission free of charge, the Agency expects to be entitled to a free of charge, non-exclusive irrevocable license to use the invention or proprietary technical data, produced by the experimenter, for their own purposes in the field of space research and technology. This old ESA rule, resulting from former scientific research programmes, is now under discussion among the Member States and has to be adjusted to the changing utilization scenario of the ISS. The up to now still existing regulation would be a severe obstacle for the involvement of commercial ISS users and its readjustment is a mandatory prerequisite. ESA has already expressed its intention that for industrial research and development by the private sector on a commercial basis adequate guarantees of confidentiality, including intellectual property rights, will be secured [Reference 3].

2.2 German Utilization Concept

The German utilization participation will be handled primarily via ESA for the Columbus Orbital Facility (COF) and the external facilities; in addition there will be bilateral projects with the other ISS partners, e.g. NASA or RKA. The different users are subdivided into three utilization branches: basic science, application oriented research and industrial/commercial utilization. Although no strict requirement, it is anticipated that each branch will take about one third of the overall German utilization resources. The ISS utilization is based on excellent science with increasing industrial participation.

2.3 New Quality of Resources

Up to now the available resources in manned spaceflight have been limited; energy, crew time, up- and download, mission frequency and duration depended on the availability of Shuttle flights or possible participation in Mir projects. As Germany does not have its own access to space, the situation was even worse here. Now a new dimension opens up — the continuous availability of all the above resources, with regular flights up and down and a permanent habitation.

Comparing the Columbus Orbital Facility with Spacelab, it is like having a permanent Spacelab mission. 8.3% of the non-Russian ISS resources and 51% of the COF resources are for European use [Reference 4]. For Germany this sums to approximately a quarter of a Spacelab mission; however, it is on a permanent basis. This is an absolutely new boundary condition which makes the utilization valuable and calculable for industrial users. Large resources, inside the modules and outside on platforms, become available on the ISS, combined with continuous access. This is a new quality which makes the ISS attractive for industrial utilization by the non-space industries.

3. Promotion of Industrial Users of the ISS

The main objective of the DLR project is to get in close contact with the non-space industries and to promote the ISS utilization there. The project systematically supports this new strategy for utilization by relevant investors. The marketing approach, which is part of the promotion project, is done outside the established space community. New multilateral contacts are established between government, organisations and companies; personal contacts are of the utmost importance. This new scenario leads to the development of internal evaluation processes among the addressed companies. We intend to stimulate an open and positive platform for decisions which may become relevant for participation in future space experiments. In parallel space solutions for terrestrial research activities will be outlined and brought to the prospective customer. The main contractor for this project is the German consulting company Kesberg, Bütfering und Partner.

The promotion project has to proceed slowly and step by step. We have to take into account still developing boundary conditions and must be very sensitive to the reactions of potential new users, creating moderate expectation levels, and act as a service provider. Because of past unsuccessful approaches to intensify industrial participation under poor access conditions, there are old prejudices against space flight participation. Although the boundary conditions are now more favourable, people do not know enough because the information is not transparent enough for them.

The DLR promotion project consists of the following major elements:

- Information Service ISS, including a quarterly ISS Newsletter and a dedicated Internet home page (www.raumstation.dlr.de)
- Workshops organised by DLR, for industries and their associations
- Working Group Innovation ISS, an experts' forum and evaluation group
- Information services for the associations of industry and Chambers of Commerce.

Besides this, several other activities like participation in exhibitions, conferences and intensive communication with potential participants from non-space industries contribute to this project which is performing successfully. The project is in very close contact with the German space industry, to exchange information about their own activities of market preparation for industrial users. This leads to an optimised effort within Germany.

4. Other Support to German Industrial Users

Besides this promotion project Germany supports at the present time two industrial users participating in the early external utilization of the ISS, and provides existing experiment hardware and facilities for use by industry.

4.1 Industrial Payloads for Early External ISS Utilization

Among the German contributions for the early external utilization of the ISS during the timeframe 2002 – 2004 are two payloads which are already mainly financed by industrial users.

Global Transmission System (GTS). This is a joint project between the German watch company Fortis, the car maker Daimler Benz and the University of Stuttgart. The intention is to provide a space-borne time signal for wrist watches with an antenna in the bracelet. Emitted via the Space Station, this signal will be available to 95% of the world's population. This is an advantage over the present radio technology, which is based on a long wave radio transmission, restricted to a 2000 km radius around the transmission source near Frankfurt/Main, Germany. Besides the time signal the project opens possibilities for world wide paging, theft protection of cars, credit cards and more. This experiment is preparing for new industrial applications using space as a means to provide a service on the ground.

High Temperature Superconductor Demonstrator for Communication Satellites (HTSC). The achievements of this technology lie in the miniaturization of key components in microwave signal amplifiers without loss of quality and reliability. These components, e.g. multiplexers and filters, operated at a high temperature superconducting state, will reduce mass and volume of space operated components and increase the signal-to-noise ratio. This experiment increases the product acceptance by system customers, prepares the market and improves the competitiveness of the component provider.

4.2 Availability of Existing Payloads for Industrial Utilization

Some other already existing space instrumentation is now being made available for industrial utilization approaches by several space industry enterprises. The commercial activities are supported by the agency in such a way that the suitable facilities are offered for use by industry for a nominal fee.

The Closed Equilibrated Biological Aquatic System (CEBAS), which was originally developed by OHB-System under contract from the German Space

Agency, is part of the Memorandum of Understanding between OHB and SPACEHAB jointly to provide commercial life science spaceflight services on the Space Shuttle and the ISS [Reference 5].

The **Commercial Protein Crystallization Facility (CPCF)**, originally developed by DASA, has been flown successfully in 1998 within the framework of an industrial experiment on the Space Shuttle and it is now anticipated to refly some of the 1998 experiments, providing a longer crystallization time. The high interest of the users is documented by their strong financial contribution. Only the provision of the flight opportunity and the mission costs is requested.

4.3 Classic Networking Combines Basic, Applied and Industrial Research

Besides these new approaches Germany continues to work on the classic network projects. On the one hand we have several national networks in the fields of materials science and life science. Other projects have been established at the European level by ESA, so called topical teams [Reference 6], or by the EU, the well known example of ISPRAM (In-situ Processing of Aluminum Matrix Composites). All of these projects are conducted with strong German participation.

5. Access Conditions and Requirements for German Industrial Users

The new user community for industrial/commercial utilization needs access conditions strongly differing from the purely scientific approach. Peer reviews are no longer a means to fulfill confidentiality requirements. Research results must be exclusively reserved for the investigating company. Other important criteria are reliable schedule and precise cost information, and the possibility for access within a very short time. These requirements are derived from the experience gained in some first industrial experiments and have been summarized at the European level by EUROSPACE [Reference 7]. With this information the potential user can calculate his risk in participating in a space experiment. Last but not least, the safety processes must be performed in a confidential manner, not disclosing experiment relevant information but still fulfilling necessary safety standards. During the ongoing preparations of the first industrial utilization projects, these conditions are evaluated and successfully applied.

Industrial proposals should be forwarded via the national space agencies and national evaluation criteria should apply. A special selection process has to be established to consider the specific industrial aspects and to assure, definitely, the required confidentiality. Priorities have to take into account the value adding and market potential of the individual proposal. The services which need to be given to industrial/commercial users will be provided first by already existing user support centres like Microgravity User Support Centre (MUSC), Microgravity Advanced Research and Support (MARS), Centre of Assistance for the Development of Microgravity and for Operations in Space (CADMOS), etc., from the former Spacelab/Mir era. New commercial service providers are growing, e.g. BEOS in Bremen. This is a new quality of service, where government is no longer the only customer. In Germany industrial/commercial users should contact DLR first; they will then get in contact with the relevant specialists for further advice and support, pending on the individual demands of the experiment.

For the time being we are in a promotion phase for ISS utilization. Thus the mission itself will be supplied without charge during the next years. Users are expected to take significant financial risk only for their own experiment activities on the ground; funding, e.g. via DLR, other national institutions or European Union is anticipated. It is clearly understood that during the next few years a funding component is mandatory to prepare the new market place in space.

6. Successes in the Past, Today and in the Future

Even with relatively counterproductive boundary conditions of the past some important and promising results for industrial enterprises could be achieved. Examples are an eye pressure measuring device, solidification of aluminum alloys used by Audi and Airbus, bearings for engines, robotics software and sensors, and protein crystallization.

The following examples of identified research topics for ongoing and future work give an idea of the broad potential behind industrial research in space: combustion processes, phase transitions in fluids, compound materials, casting and directional solidification, undercooling, amorphous solidification and glasses, thermophysical properties, crystallization of semiconductors and proteins, aerosols, colloids, and growth of three-dimensional tissue.

The current terrestrial research topics like health, biotechnology, environment, mobility and new materials are well met by the above key words. The potential benefits of the new research opportunities on the ISS have been clearly understood during the first discussions with non-space industry. Although the number of direct meetings with these potential new users was limited because of the recent start of the promotion project (autumn 1998), we have experienced a very positive resonance. Industry is open to listen to the offer and is requesting more information regarding their individual research topics. This service is provided to them within the infrastructure of the project.

7. Conclusions and Outlook

The access conditions have to be defined to comply with the needs of industrial and commercial users. Germany acts according to the above mentioned requirements within the relevant ESA boards. On a clear and transparent basis industry is expected to make the best use of the new large scale research facility in space, on the same foundation as similar facilities on the ground. Potential and promising research fields have been identified in the framework of analysis and discussion together with industry. Even with the rare access occasions of the past and today, the first successful experiments for industrial/commercial users were performed. It is our conviction that we will experience a growing market in the future.

References
1. European Space Agency: Draft Declaration on the European Participation in the International Space Station Exploitation Programme, ESA/PB-MS(99)8, March 5, 1999
2. European Space Agency: Programme Proposal for the European Participation in the Exploitation of the International Space Station, ESA/PB-MS(99)7,rev.1, March 23, 1999
3. European Space Agency: The International Space Station, A Tool For Industrial Research, ESA BR-136, October 1998
4. European Space Agency: The International Space Station, A Guide For European Users, ESA BR-137, February 1999
5. OHB-System GMBH and SPACEHAB, Inc.: OHB and SPACEHAB agree to develop commercial experiment facility for the ISS, http://www.spacehab.com/press/99_03_31.htm, April 13, 1999
6. European Space Agency: Approval of the Topical Teams in Physical Sciences, ESA/PB-MG(96)3, February 1999
7. EUROSPACE Report: Industrial Utilisation of the ISS by European Space Industry, January 1999

Commercial Management for the Space Station: Making the ISS More Accessible to All

G. Inoue[1], Inmarsat Ltd., 99 City Road, London, EC1Y 1AX, UK

e-mail: Detto@aol.com

J. Maroothynaden, Imperial College of Science, Technology and Medicine, UK

e-mail: j.maroothynaden@ic.ac.uk

Abstract
    The International Space Station represents a remarkable milestone: if managed correctly, it has the potential of becoming the first truly global marketplace, one set above all national borders. Yet, contrary to its name, the ISS is currently far from being international. Under current agreements, it is, in fact, only accessible through the national space agencies of a select group of participating nations.
    In this paper, an innovative concept is suggested where the use of the ISS is managed in a fair and efficient manner, allowing not only the ISS participating nations but all users, including developing nations and commercial companies, to benefit from use of this international orbiting station. The proposal is based upon the creation of a 'quasi-IGO', whose main business would be to run and manage efficiently the allocation of Space Station resources to potential users. The company would be set up by a consortium of international businesses and would act as a single point contact between the end user customers and the national space agencies of the ISS participating nations. The Company would also offer other services.
    A concept of a 'participation fee' is introduced in the paper, whereby a fee is collected from all parties wishing to send payloads to the ISS, except from the initial consortium of founding companies. The collected funds are used to subsidise experiments from educational users and those from the developing countries, thereby making the Space Station more accessible to all. In addition to the fees, users are also subject to a charge for each payload or experiment they wish to send, which would vary according to the launch and operation costs, and the amount of resources required aboard the station. As a conclusion, the current trend of commercial businesses being encouraged to access space without having recourse to public funds is mentioned, and the potentially attractive implications which this has on the proposed venture is explained.

1. The International Space Station

    The International Space Station (ISS) will be the most technically advanced, permanently inhabited man-made structure to orbit the Earth. Its construction has been achieved through Intergovernmental Agreements (IGA) between 7 national partners (United States, Russia, Canada, Japan, ESA, Italy[2] and Brazil) resulting in the design, manufacture and delivery of various flight components

---

[1] George Inoue currently works as a finance officer at Inmarsat Ltd. The views presented in this paper are those of the author and do not reflect those of the organisation.
[2] Although part of the European partner, Italy is providing the American designed temporary module - ASI Multi Purpose Logistics Module.

G. Haskell and M. Rycroft (eds.), International Space Station, 163-174.
© 2000 *Kluwer Academic Publishers.*

and associated ground support elements [References 1,2]. The US National Aeronautics and Space Administration (NASA) is currently leading the ISS development and implementation program by providing the main elements and modules, in conjunction with the Russian Space Agency (RSA).

2.   Utilisation of the International Space Station

2.1   Users

Potential uses of the ISS may be characterised into 4 distinct types - academic, industrial, media and educational. Together they cater for a wide range of applications (Table 1).

Academic uses may also be combined with industrial users via research collaborations, where the emphasis is on the development/qualification of hardware and products for commercial gain (i.e. pharmaceuticals, or electronic chips through microgravity research). Media and educational utilisation use spaceflight in the form of films, educational programs, advertising and exhibitions to educate or entertain the public. Commercial emphasis is exhibited by media utilisation due to advertisements, product placements and selling educational material.

Applications	Academic	Industrial	Media	Education
Natural sciences	X			
Engineering	X	X		
Manufacturing	X	X		
Instrumentation/detection	X	X		
On-orbit crew utilities	X	X		
Remote Sensing	X	X		
Test bed / spaceflight qualification	X	X	X	
Education	X		X	X
Entertainment	X	X	X	X

Table 1. Current spaceflight utilisation

2.2   Getting Access to Space

Currently, all potential users are required to undergo a mechanism consisting of proposal submission, evaluation and selection in order to get access to space. The stages involved in the experiment proposal selection are usually defined by the respective agencies of each partner nation but are presented here in their general form (Table 2). There are 7 stages that may be divided into three phases: initial, core and final.

In the initial phase, an Announcement of Opportunity (AO) is issued by the respective space agency, and proposals for this are collected from potential users. Industrial users have the option of 'by-passing' the time consuming AO stage by paying a participation fee, but are still mandated to go through the core selection stage. The core selection phase is also time consuming, consisting of a review by peer panels on the scientific and technical merit of the proposal. Once the proposal is accepted as technically suitable for flight, it is reviewed again to determine if it is relevant to the research and development programs of the various space agencies.

Stages	Academic	Industrial	Media	General Public	Time
Agency announcement (AO)	X		X	X	3 month
Proposal submission	X	X	X	X	
Scientific and technical merit evaluation	X	X	X	X	5 months
Assessment of technical feasibility	X	X	X	X	
Review of relevance to agency	X	X	X	X	
Ground based R&D	X	X		X	> 4 years
Flight	X	X	X	X	
Post-flight	X	X	X	X	

Table 2. Current proposal selection procedure

As can be seen in Table 2, the time from proposal submission to flight is currently very long (greater than 4 years). This is due to the ground based R&D phase where the proposal is transformed from a paper concept to flight hardware and protocols. This time lag may be reduced (by 1 or 2 years) by industrial and media users who often use 'off the shelf' space-qualified products in the experiment designs.

2.3   Who has Access?

The United States and Russia currently have the most regular and reliable fleet of launchers to access orbiting platforms. Through the IGAs, NASA and the RSA allocate a number of flight opportunities, which are then bartered for by the partner states. The flight opportunities are then advertised via AOs to nationals stipulated by the advertising partner agency. It is then up to the partner agency to allocate flight opportunities to other, non-participating nations, depending on their levels of contributions to that agency.

2.4   Problems with the Current System

According to the IGA, NASA has the lead role in the ISS utilisation program.   The IGA require the amount of utilisation, operation and management controls delegated to each partner nation to be proportional to their contribution levels.   In addition to this, partners are fully responsible for maintaining and developing their own ISS elements, and the common ISS operational costs are to be spread amongst them.   (For example, ESA has been assigned 51% usage of its own Columbus module and 8.3% usage of ISS on-orbit utilisation resources once the ISS is fully operational [Reference 1].)

The resultant effect is that, for the first few years, the United States will have the majority share of ISS resources, and partner payloads may not be flown until at least 2001.   This means that, whilst the partners are contributing to the ISS, their experiments and crew members will not be able to reap the benefits until around 2003, when it is expected partner payloads will be fully integrated within the Space Station [Reference 2].

Another major problem is the political influence in the experiment selection procedure (which has to go through national space agencies).   This inhibits a number of users, especially commercial, from having their proposals selected for flight due to the nature of the payloads.   Furthermore, these space agencies would be more inclined to approve politically attractive payloads that are of an educational or scientific nature, for example, as they are more in line with the public perception of what national space agencies should be doing.   This political aspect of the agencies cannot be avoided, as they are essentially government organisations, and have to account for their budgets to taxpayers.

A possible solution is proposed to ease this process and allow easier, and timely, access to space, both of which are fundamentally important if we are to increase the use of the ISS by commercial users.

3.   Solution: A Commercial Company to Manage Commercial ISS Payloads

The majority of problems mentioned in the previous section may be eliminated by having a commercial company act as a 'manager'[3] of commercial proposals for the ISS.

The main service provided by the Company is to act as an international single point of contact to allow all user's (Table 3), irrespective of nationality, to

---

[3] The Company will only manage payload volume onboard ISS that has been allocated for commercial purposes only.

benefit from easy access to the International Space Station for their payloads. To do this, the Company is to provide all the services that may be required when sending payloads to the ISS. These are explained in the following sections.

Established Applications	New/Imaginative Applications
Natural sciences	Crew training
Engineering	Mission operations (ground and in-flight)
Manufacturing	Repair and maintenance (ground and in-flight)
Instrumentation/detection	Crew and cargo delivery/return systems
On-orbit crew utilities	Experiment design support
Remote sensing	Tourism
Test bed / spaceflight qualification	Free flyers
Education	Communications
Entertainment	Transportation device development
	Technology development

Table 3. Current and potential flight user applications for ISS

3.1 Services Provided by the Company

Management of payload proposals. The main service provided by the Company is to manage commercial proposals of ISS payloads. The process of screening proposed payloads is extremely complicated and is briefly dealt with in this paper. The first step involves the separation between payloads whose owners are prepared to pay a premium in exchange for a swifter launch, and those who would prefer a cheaper launch, despite longer waits. Secondly, the proposal would go through the techical feasibility check process (Table 2). The third step would be to collect similar payloads and try to match them onto a single mission.

In addition, the Company would also be responsible for ensuring that the various safety, import/export regulations for the launch vehicle state are strictly adhered to by the payloads, prior to the shipping to the launch site.

ISS resource management. The Company would be responsible for ensuring that the required resources onboard the ISS are available. This would be mostly done though negotiations with the participating nations, by purchasing ISS volume through the respective module owners, and either re-selling or leasing it out to the proposal managers. The idea is that the Company would have more weight in negotiations with participating nations to acquire the resources than individual users (through professional contacts and frequent dealings).

Specialist advice / Consultant services. Another service to be provided by the company involves giving specialist advice on the various international regulations concerning the import/export of high-tech payloads, radio frequency issues, and intellectual property issues on board the ISS, among others. In order to provide such services, the Company would strive to employ experienced consultants in relevant fields, in addition to having special contacts within the various organisations (such as NASA, US Government) where advice may be obtained directly from the policy makers.

The Company is also to offer assistance to users (especially those from the developing nations) to identify additional sources of potential funding.

Competitive insurance premiums. The Company may also offer competitive premiums for insuring payloads through negotiations with global insurance houses, and various space agencies on behalf of the users. Once established, the Company may also act as an ´insurance broker´, offering users a choice of services from favourable insurance companies.

Cheaper launch prices. The Company may be in a position to negotiate mass bookings of launch space aboard a portfolio of launch vehicles. By booking in bulk, cheaper launch prices may be negotiated. Also, similar payloads will be combined and flown on dedicated missions. These measures will result in lower prices for the user.

3.2 Structure of the Company

In-order to realise this company, we propose a unique 2-stage process. The first is to create an amalgam of an Intergovernmental Organisation (IGO) and a commercial corporation to enable the creation of the Company. The second stage is to transform the ´quasi - IGO´ to a profit making commercial company, independent from poltical influences.

IGOs and their structures. Historically IGOs were mainly created in situations where the participation and backing of a number of governments was necessary to allow the project or system in question to operate. One good example of this is in the satellite telecommunications industry where a number of IGOs were created in the 1970's. By having a number of governments participate in and fund the IGO, the business and financial risks to each individual nation were decreased, whilst at the same time allowing the nations to benefit from the potentials offered by satellite based telecommunications.

The structures of the various intergovernmental organisations typically consist of a number of bodies, usually representing the organisation workforce, the participating governments and the companies with vested interests. The structure of one such IGO, the International Mobile Satellite Organisation (INMARSAT) is a good example (for more information, refer to Reference 3.)

Problems with the IGO structure. One of the main drawbacks of the IGO structure is in its inability to adapt quickly to new business demands to remain competitive. As a result, it is now widely recognised that such organisational structures may only be effective as long as there are no commercial competitors providing similar services.

In the case of the satellite telecommunications industry, the market has developed enough for the private sector to become more willing to bear the risks of starting up private satellite based ventures. As such, a number of commercial companies offering similar services have been created since the advent of the IGOs. These companies, being commercial, are better able to adapt to the changing market demands than the intergovernmental organisations, and over the years these companies have been 'eating away' at the market, previously dominated by the IGOs. As such, a number of IGOs have recently announced their intent to dissolve and privatise into commercial entities in order to remain competitive - Eutelsat, Intelsat [References 4, 5], and one, Inmarsat, has recently become the first IGO successfully to undergo privatisation on April 15, 1999.

Quasi-IGO: The first stage for the proposed company. For our proposed company, it is clear that the market demand for such services as management of ISS payloads has not yet developed. It is also clear that no single government or private entity is prepared to bear the risk of providing these services. It may therefore be said that the situation is similar to the conditions where IGOs were initially created, and hence one may be tempted to propose an IGO structure for the Company. However, to avoid the problems facing IGOs once the market develops, it is suggested that a new, innovative organisation be created for this Company.

A 'quasi-IGO' structure consisting of a consortium of governmental agencies, corporate entities and other organisations that is operated and managed by a board of directors, typical of a private commercial company, is proposed. This consortium will most likely be composed of entities which stand to make frequent use of the company's services due to the various privileges and concessions granted to them, once the company becomes operational. Examples of entities that might be interested in forming the consortium include

NASA and ESA as governmental agencies, large pharmaceutical companies for corporate entities, and universities and research institutions.

Implementation of 'quasi - IGO'. The consortium is to be responsible for providing the initial financing to set-up and operate the Company, whilst the board of directors will be responsible for the daily operation of the Company. The initial board of directors of the new Company is to be elected by the consortium, and most probably be representative of the investment share of each consortium entity, very similar to an IGO.

As the Company is to be fully commercial, it must be stressed that ALL management decisions regarding its operation remain with the company board of directors, and NOT the consortium. They will get a fair return on their investment through a waiver of participation fees, and through other benefits as described in the following sections. This is essential to allow the Company's board of directors to act in the best interest of the company, and not of that of the consortium. Additionally, by giving complete management autonomy to the board, the Company will effectively be able to select the ISS payloads impartially without prejudice or political preference to payloads originating from member states of the consortium.

Commercial company: The second stage. As the market develops and private industry begins to warm up to the high potentials presented by the ISS, the Company may be able to secure additional funding through the usual means of issuing debt and, at the appropriate time, floating on the stock markets through an IPO (Initial Public Offering). Once this is achieved, the role of the consortium as financiers of the venture will change, but they will still retain their various benefits as founding members. As per the norm, the shareholder will then be able to elect new members to the board of directors in the usual corporate manner.

Hence, the proposed 2-stage structure would allow the company to have a truly international personality, without having any governmental involvement, thereby allowing the company to maintain its competitive edge, and also facilitating the payload selection process in a truly impartial manner.

3.3   Participation Fee

All users wishing to send a payload to the ISS through the company will be required to pay a 'participation fee'. This is somewhat akin to the ESA membership structure [Reference 6]. This fee will reflect the individual usage of the company's services and is to be levied on all users, except 'special' members, and those members of the consortium. (As it is most likely that members of the consortium are the highest users of the services provided by the company, they will stand to benefit the most from this concession of the participation fee.) The main reason for the participation fee is to cover various operational costs for the Company, as well as to provide for a mechanism to enable partial funding of payloads from developing nations and other, non-profit organisations, by the Company. It is hoped that such a policy would not only allow for truly global access to the ISS, but also help justify to governmental agencies that the Company is worth the investment.

A tiered charging system. As mentioned previously, the fee should be levied to each participant reflecting the various levels of ISS utilisation, on a yearly basis. For regular ISS users (i.e. at least one flight every three years), the nominal annual participation fee would be appropriate. This nominal fee is to be paid every year, irrespective of whether a payload is scheduled for launch in that year or not.

For 'one-off' or irregular users, the participation fee would be correspondingly higher than the nominal fee, yet will only be applicable to the calendar year in which the payload is to be considered for launch. This 'higher rate' will be priced in such a way as to make it more economical for regular users to continue paying for the nominal rate on a yearly basis, as long as on average they intend to send a payload into space once every three years.

It is hoped that this participation fee structure would avoid the problem of 'one-off' or irregular users benefiting unfairly from the annual fees paid by the regular users.

Special concessions. A specially reduced participation fee would be offered to non-profit organisations such as universities and to developing nations. This would act as an incentive to send payloads to ISS. For academic institutions, a condition for the discounted fee is that the results obtained by the ISS flown experiments are made publicly available to all participating members (including the consortium). The level of concession would vary on a case-by-case basis, and would depend on the amount of pooled funds available from the participation fees paid by others.

3.4. Pricing Issues

The participation fee described in the previous section does not include any payload management, launch or insurance costs. The cost for each payload is to be evaluated individually, and this will depend on a number of factors. These include:

- Placement within mission depending on the complexity of the payload. Standard, and common, payloads (such as telecommunications) will obviously be easier to place within similar missions, and hence would incur less administrative cost than 'unusual' payloads
- The timeline needed for payload launch. The longer the time to launch, the lower the cost. This is because, with additional time, the company will be more able to wait for an appropriate mission with which to include the payload
- Payloads requiring 'extra care', such as cryogenic cooling, or extra testing and verification to ensure flight safety, would incur extra charges accordingly
- Payload dimensions. This is because the physical size and weight of the payload would play a crucial role in determining the launch vehicle and whether or not it is possible to include it as a 'piggy-back' launch with other payloads
- Selection of the launch vehicle for their payloads, from a list provided to them by the Company including the choice of 'piggy-back' rides or dedicated launches; insurance premiums; import/export licences and additional legal requirements.

3.5 Financing of Member Payloads

To remove bias, it is suggested that ALL payloads should be priced using the same methods, irrespective of the member status. Once the costs have been determined, negotiations should then take place to determine the amount of financing which the member payload should receive, depending on its status (i.e. from developing nation, or non-profit organisation) and needs. The size of the fund available at that time should also be considered. As previously mentioned, this fund is to be gathered from the participation fees and from other sponsorship plans.

The Company is to provide minimal funding only, with the main emphasis being on helping the users to find external sources of funding though industrial and governmental contacts.

3.6 Main Risks to the Company

Market risks. The proposed Company is especially sensitive to market risks, as the main business of the Company does not yet have a developed market. Indeed, one may even consider the business to be more of a market 'driver' rather than being market led. It is actually anticipated that the market demand for payload access aboard the ISS will develop further, once the company has been established.

The risks are best reduced through extensive market research to establish accurate forecast models to ensure that the demand for such services, and the prices which the market is willing to tolerate, are properly determined and accounted for. Recent studies have concluded that the market for such services will indeed flourish, especially from the private sector [Reference 7].

Recent legislation passed by the US Congress mandating NASA to use commercial services for space activities, wherever possible, has also helped to reduce concerns over the development of this market [References 8, 9, 10]. Such legislation would also ensure that ISS payload customers applying for access via NASA would more likely be transferred back to the proposed commercial Company by the agency. Alternatively, payloads may still be managed by NASA, but the proposed Company would act as a subcontractor to NASA.

Financial risks. Financial risks are a major hurdle for the Company. The initial financial requirements are quite substantial, but not unreasonable, especially compared to the amount of funding raised by other 'start-up' ventures in the satellite telecommunications industry [References 11]. Whilst the business case for mobile satellite communications ventures may probably be more obvious to investors than that of payload access aboard the ISS, the inherent business and financial risks are very similar. Both are very high risk, and both aim to address a new market, one that has yet to be developed. As these mobile satellite ventures have managed to secure the funding from the banks and private investors, it may be reasonable to assume that the proposed Company should be able to manage the same.

Legal risks. Legal risks are also significant. The ISS has no legal precedent. It is the first time that the world has created an orbiting platform, composed of modules manufactured by different nations. To allow commercial payloads from foreign nations to reside and operate in these modules would indeed raise interesting legal issues. Furthermore, significant legal developments are also required to allow private companies to send and manage ISS payloads. One of the main purposes of including governmental agencies such as NASA and ESA in the initial consortium was to ensure that, through their active involvement, governments would be more inclined to push for legislation reforms, to facilitate the implementation and operations of the Company.

4.   Conclusion

The International Space Station presents a unique opportunity for global users to benefit from having an orbiting platform to host payloads over long periods for various purposes. Due mostly to political issues, the current procedure for gaining access to the Space Station is not very favourable to users, especially for those not part of the ISS participating nations. In order to overcome these problems, an innovative concept has been proposed in this paper, where a private commercial Company is to run and manage the payload selection procedure, and all other phases including up to launch of the payloads to the ISS. The proposed Company faces a number of business and financial risks, the main one being the possibility that the market for such services (providing easy access to the ISS) fails to develop swiftly enough. However, despite the risks, it has been concluded that the potential benefits from the venture would be greater, making the investment worthwhile. In addition, recently passed legislation in the US has paved the way in allowing such commercial space activities to develop, and it is anticipated that other space-faring nations will soon follow suit in implementing similar legislation.

It should be noted that, whilst there are a large number of political, financial and business hurdles to overcome in implementing the Company, the message is clear: if the ISS is to become a prime tool in promoting and developing commercial activities in space, the current system for ISS access will have to change. The proposed Company is an example of how to do this.

References
1.   ESA Space Station Utlisation Division: *The International Space Station - European Users Guide.* Directorate of Manned Spaceflight and Microgravity, ESTEC, Noordwijk, The Netherlands, 1998
2.   Bartoe, J. D. F.: *International Space Station - Programs and Research Prospectus,* Presented at the Second European Symposium for the Utilisation of the ISS, Paris, ESTEC, Noordwijk, The Netherlands, November 16, 1998
3.   Inmarsat Ltd.: *Inmarsat Internet Site,* <http://www.inmarsat.org>.   Inmarsat, London, UK, April 4, 1999
4.   Eutelsat: *Eutelsat Internet Site,* <http://www.eutelsat.com>. Eutelsat, Paris, France. April 10, 1999
5.   Intelsat: *International Telecommunications Satellite Organisation (Intelsat) Internet Site,* <http://www.intelsat.int>. Intelsat, Washington D.C., United States, April 8, 1999
6.   ESTEC Public Relations Office: *ESA.* November 1995
7.   Personal notes of ISU Lectures by industry officials, January-February 1998
8.   NASA Act of 1999 (Introduced in House), HR. 1654 SEC.203 *Commercial Space Goods and Services,* <http://thomas.loc.gov>. May 2, 1999
9.   Commercial Space Competitiveness Act of 1999 (Introduced in House), HR. 1526, <http://thomas.loc.gov>. May 2, 1999
10. Commercial Space Act of 1998 (Agreed by the House), H.RES.572, <http://thomas.loc.gov>. May 2, 1999
11. Iridium: *Iridium LLC. Internet Site,* <http://www.iridium.com>. Iridium, Reston, VA, United States, April 20, 1999

A United Nations Module on ISS: A Study

V. **Lappas**, International Space University, Strasbourg Central Campus, Parc d'Innovation, Boulevard Gonthier d'Andernach, 67400 Illkirch-Graffenstaden, France

e-mail: lappas@mss.isunet.edu

Abstract
In November 1998, the space community saw the beginning of a new era, the era of the International Space Station. It is an effort that materializes the effort of many nations to cooperate and to establish a permanent manned presence in space. But is this effort truly international, or rather multi-national? Are there any opportunities for truly international cooperation for the ISS? This paper tries to answer these questions by describing a new concept on how to bring vision, international cooperation and reality within the ISS framework.

This idea involves using existing technologies/hardware on Mir to build an international environment on ISS under the umbrella of the United Nations. This will enable countries which are members of the UN, but which do not have direct, autonomous access to space (i.e. not major space powers) to utilize a proposed UN module for scientific research, space-related projects or experiments. In this paper, the relevant technical and political issues are addressed, resulting in a proposal involving more partners/participants on ISS.

1. Introduction

On the verge of the second Space Shuttle mission to the International Space Station (ISS), the space community has finally realized that in one way or another ISS is a reality. Even though ISS initially began as a one nation project (Freedom, USA), it eventually evolved to a Western multi-nation project including Japan, Europe and Canada. This plan changed after the end of the Cold War with the addition of Russia, giving ISS an international flavor.

Many things have changed in space. Today, there is a renaissance in the space community in all areas. New launchers, new business ventures and more and more countries without previous experience are either getting (or planning to get) involved in space related activities. Although heavily and continuously criticized by all, ISS is the flagship of this renaissance in space. One of the most interesting aspects of ISS is the interest of a number of non-ISS partner countries to conduct research on ISS.

2. New Cooperating Partners-Users on ISS

There are many countries interested in becoming involved on the ISS. These countries members of the former Eastern Pact, gained useful experience on Salyut and Mir Space Stations. These countries, such as Bulgaria, Poland, Czech Republic and many others (e.g. India, Australia) had very interesting space programs and have accumulated much experience and expertise. Many of

G. Haskell and M. Rycroft (eds.), International Space Station, 175-180.

them also had flight opportunities to space as well. Thus it is natural that these countries are seeking cost effective avenues to become involved with ISS. There are also many other countries that have the potential of becoming cooperating states such as South Africa, Australia, Chile, China and Ukraine. But it is very difficult for all these countries, when more research and science have to be done with limited or decreasing national funds.

Thus, an organization that would concentrate, analyze and materialize various proposals which could be made by these countries in the near future, in a cost effective manner, is needed.

2.1 The United Nations Committee on the Peaceful Uses of Outer Space (UNCOPUOS)

The General Assembly, in its resolution 37/90 of 10 December 1982, decided, upon the recommendation of the Second United Nations Conference on the Exploration and Peaceful Uses of Outer Space (UNISPACE), that the United Nations Program on Space Applications, *inter alia*, should promote greater cooperation in space science and technology between developed and developing countries, as well as among developing countries [Reference 1].

The UNCOPUOS public outreach programs aim to support, develop and sustain the direct participation of developing countries in front-line activities, in a three-phase approach involving:

- Basic space science education
- Further development of locally (and regionally) identified research and educational facilities, such as networked modern astronomical observatories of moderate size
- Direct access to facilities for front-line basic space science

The UNCOPUOS outreach program provides an excellent platform to conduct international, multidisciplinary research on ISS between developing and non-developing countries. There are interesting examples from the UNICOPUOS outreach program that prove this point, such as the proposed World Space Observatory, the Sri Lanka Telescope Facility, the Central American Astronomical Observatory in Honduras, and many others [Reference 2].

2.2 A United Nations Module on the International Space Station

Although ambitious as a concept, the idea of developing countries as well as others being able to conduct space-related research on a facility in space is too intriguing to ignore. What better facility could possibly exist than ISS to promote space science education, further develop research and provide access to a first-class, front-line facility? The advantages of a UN facility on ISS would be to:

- Stimulate space related research in developing countries
- Provide flight opportunities for countries which are members of the UN in cooperation with ISS partners
- Promote greater cooperation in space science and technology between industrialized and developing countries, as well as among developing countries
- Gain further public support for ISS
- Accelerate ISS utilization
- Have a truly international ISS
- Complement, or become the space segment of, the proposed World Space Observatory
- Stimulate the creation of a single, international coordination and decision making body for ISS operations and utilization, based on the contribution of each partner.

Instead of having to build a costly new module US $ 300-400 million, the example of the Sri Lanka telescope could be used, where Japan donated the necessary equipment to Sri Lanka. It is proposed that just before the de-orbit of the Mir Space Station (whenever this might be planned) one of its existing modules could be transferred to ISS.

The two possible candidates for a UN module from Mir would be the Krystall module and the newer Priroda module (see Fig. 1). The choice of which module to select would be based on the research to be conducted by the interested parties — Earth observations, or space science and materials science. Other factors affecting the choice are the technical characteristics of the modules and their age [Reference 3].

2.3 The Kristall Module

The Former Soviet Union added the Krystall module to Mir in June 1990. The module contains experiment space inside for biological and materials science experiments. It also hosts solar panels ($72m^2$ total area) that provided up to 8.4 kW; these can be folded or unfolded as a function of electrical power requirements to generate power for Mir. A 360 A-hr NiCd battery system provides energy storage. The module also comes equipped with a special "androgynous docking mechanism" at the far end of the module. This mechanism is used for docking with the Space Shuttle.

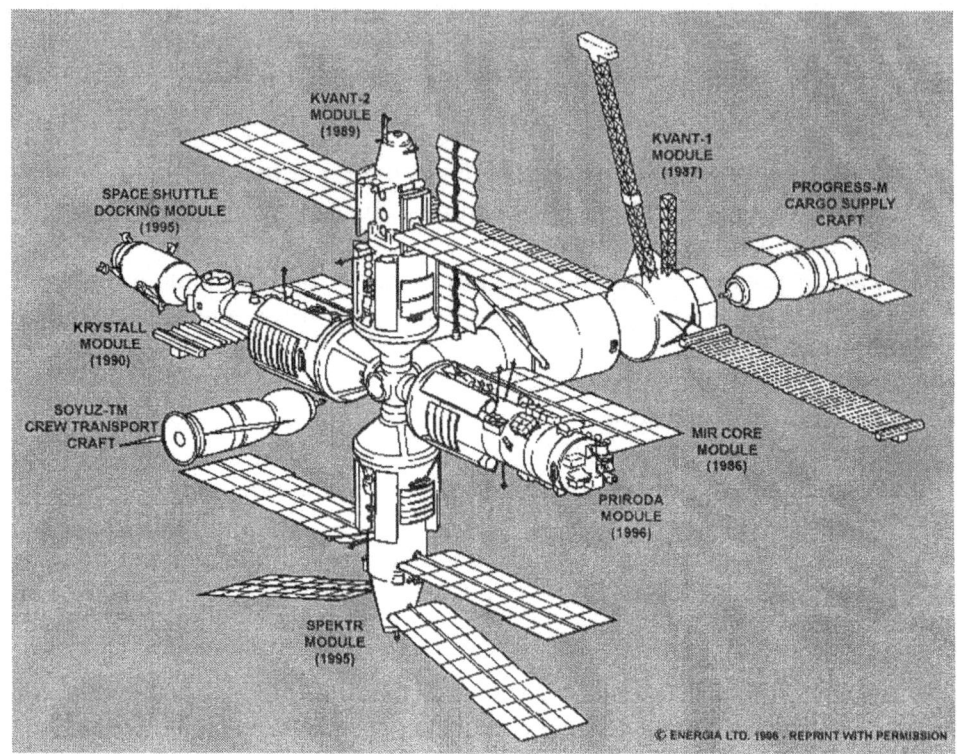

Figure 1. Mir Space Station with Priroda and Krystall modules

The instruments carried are as follows:

* Krater 5, Optizon 1, CSK-1/Kristallizator semiconductor materials processing furnaces
* Zona 2/3 materials processing furnaces
* Glazar 2 UV telescope -cosmic radiation studies
* Earth resources camera system-2 KFA-1000 film cameras
* Svet plant cultivation unit
* Mariya magnetic spectrometer
* Marina gamma ray telescope
* Buket gamma ray spectrometer
* Granar astrophysics spectrometer
* Ainur electrophoresis unit.

2.4 The Priroda Module

The most recent addition to Mir, the Priroda ("Nature") module, was launched in April 1996, completing the assembly of the Mir complex. The module carries Earth observing equipment as well as experiments and other equipment used in the joint American-Russian missions on Mir. There are several purposes for the Earth remote sensing mission of Priroda. It is designed to study the atmosphere and oceans, with an emphasis on pollution and other environmental impacts of human activities. It is also designed to conduct geological surveys that can be used to locate mineral resources and water reserves, and study the effects of erosion on crops and forests. It is also designed to receive and relay information from "emergency buoys" located in seismically active areas, around nuclear power plants, and other zones, as part of the Kentavr monitoring and warning system.

The instruments aboard are as follows:

- Ikar passive and active microwave polarization radiometers
- Travers: 2 frequency SAR
- Istok-1: 64 channel infrared radiometer for ocean research
- Ozon M: spectrometer to measure ozone and aerosol concentrations
- MOZ-Ozbor: 17 channel spectrometer to measure reflected solar radiation
- MSU-SK: medium resolution MSU-E high resolution scanners
- Centaur: Geophysical station interrogation
- Alisar: lidar aerosol.

2.5 Technical Issues

The first two issues that arise with the idea of utilizing a Mir module for ISS are transportation and compatibility. For transportation, by utilizing a simple Hohmann combined plane change (assuming that both stations are at about the same inclination and waiting for their planes to coincide) from 350 km to 450 km, the ΔV required is in the range of 0.4 km/s. A modified Progress M would be able to perform such a task, as to transfer one of the modules to ISS. As for compatibility the Russian segment of ISS uses the same docking and interconnect techniques as in Mir. The modules already have solar arrays, which would be of degraded capability and a replacement of those would be necessary in the future. Another issue is where in the Russian segment one of these modules should be placed. Depending on the module, the docking position should be carefully selected in order to ensure there will be no obstacles to observe the Earth or space or, if the Krystall module is selected, to ensure that the Space Shuttle can dock to it.

3. Conclusions

As in all space projects, the first issue that arises is that of cost. It is understandable that, in these times when the space community is trying to find ways to commercialize its projects, proposing a project with small commercial potential in an already very expensive project will be viewed with criticism even if there are ways to minimize costs (donation of module, transfer vehicle, shared costs between interested parties, rent facilities and space). It is, however, very compelling to see how such a project could become a truly international bridge of cooperation and of sustained development. ISS, from its early stages has shown to us that a new way of thinking is needed to be used in the technical, business and commercialization fields related to ISS. Maybe a United Nations initiative (module) might be able to stimulate all partners to think differently. It might also convince all partners to adapt a consolidated, international and interdisciplinary ISS strategy with input from all interested parties such as government, academia and industry that could also become the precursor to a unified and consolidated ISS coordination body that would include functions such as mission planning, utilization and operations. A useful parallel that could be used is that of the International Space University (ISU) and ISS. ISU has clearly proven that, no matter where people come from, what their colour or specific education is, as long as they strongly believe in the same goals and work hard, a lot can be achieved. ISS should provide the opportunity to all nations, whether they are space or non-space powers, developing or non-developing countries, to promote space science and technology, education and space business. Then, ISS would truly be the International Space Station.

References
1.   Haubold, H.J.: Worldwide Development of Astronomy: The story of a decade of UN/ESA workshops on basic space science, *Space Technology, Vol.18*, pp. 149-156, 1998
2.   United Nations, Committee on the Peaceful Uses of Outer Space: Report on the seventh United Nations/European Space Agency Workshop on basic space science: Small Astronomical Telescopes and Satellites in education and research, hosted by the Observatorio Astronomico De La Universidad Nacional Autonoma De Honduras, on behalf of the government of Honduras, Report number A/AC.105/682, January, 1998
3.   NASA Shuttle-Mir Web: Phase I program, http://shuttle-mir.nasa.gov/ops/mir. May 5, 1999

Report on Panel Discussion 4

Innovative Approaches to Legal and Regulatory Issues

O. Tomofumi, R. Mittal, International Space University, Strasbourg Central Campus, Parc d'Innovation, Boulevard Gonthier d'Andernach, 67400 Illkirch-Graffenstaden, France

e-mail: tomofumi@mss.isunet.edu, mittal@mss.isunet.edu

Panel Chair: D. Rausch, NASA, USA

Panel Members:

F. Claasen, German Aerospace Center, DLR, Germany
A. Eddy, CSA, Canada
A. Farand, ESA
J. Richardson, Potomac Institute for Policy Studies, USA

The first and main question proposed was: "would it be useful, in the current legal regime and political scenario, to use the UN system and its specialised agencies (like COPUOS, International Civil Aviation Organisation, etc.) to promote ISS commercialization further?"

A. Farand: In the 1960's and 70's, the UN was in the forefront of making rules for activities in space. Then individual states took over from the UN. The nature of IGAs and other instruments is such that negotiations between 5 or more partners is difficult and would be even more difficult if more countries are involved through the UN system. But there is still the need for the UN to play a role.

A. Eddy: The Canadian Space Agency (CSA) is not averse to the UN promoting the commercialisation of the ISS. CSA has already been approached by many non partner countries for access to and use of the ISS, and looks forward to a very meaningful role to be played by the UN.

F. Claasen: It is very difficult for ESA to reach a single opinion on behalf of its 15 members; how much more difficult it would be for all the countries to reach a consensus under the UN. Moreover, ESA is open to every country's participation in the ISS utilisation programme under the present IGA structure.

J. Richardson: Our world is "idea rich"; under the UN, we can come up with many ideas and the partners can debate these and decide.

G. Haskell and M. Rycroft (eds.), International Space Station, 181-182.

Another question was: "in the event of an ISS payload obtaining remote sensing data for military applications, would the ISS become a military target? And in that case would it be subject to the rules of war?" **J. Richardson** did not see any difference between the ISS and any other civilian place becoming a military target. **A. Farand** noted that the IGA requires the ISS to be used always for peaceful purposes.

What do the panelists, as government agencies' representatives, expect from users in private industry? **A. Eddy's** answer was to "bring money". **J. Richardson's** view was that the public and private sectors have to work in tandem.

Session 5

Technical and Management Innovations for ISS

Session Chair:

M. Uhran, NASA, USA

Commercial Development of the International Space Station

M. L. Uhran, Space Station Utilization, National Aeronautics and Space Administration, Code US, 300 E Street SW, Washington, DC, 20546-0001, USA

e-mail: muhran@hq.nasa.gov

Abstract
The Commercial Space Act of 1998 was signed into law by President William Clinton on 21 October 1998 (Public Law 105-303). Section 101 of the Act addresses Commercialization of the Space Station and establishes the economic development of Earth orbital space as a priority goal of the space station, while encouraging the fullest possible engagement of commercial providers and users in order to reduce government operating costs. The Act also requires NASA to submit to the Congress a series of reports delineating potential commercial opportunities, specific policies and initiatives to stimulate economic development, and an independent market study.
In response to the law, NASA has produced a Commercial Development Plan for the International Space Station which proposes a range of actions to be undertaken in pursuit of the Legislative and Executive Branch objectives. These actions fall into three broad categories: (1) completion of an internal study on potential pathfinders for economic development, to be complemented by the independent external market study; (2) approval of an Organizational Work Instruction (OWI), under the auspices of the agency-wide ISO-9000 certification initiative, to establish a clearing house function at NASA headquarters for the dispositioning and auditing of commercial proposals; and (3) development of the concept for a Non-Government Organization (NGO) to manage utilization and economic development of the United States stake in the International Space Station.
These actions are currently in the nascent stage; however, they are anticipated to grow in terms of their impact on space station utilization and operations as the assembly sequence and economic development plan progress.

1. The International Space Station and Microgravity

Just six months ago, the Russian *Zarya* spacecraft and the United States *Unity* node were successfully joined in orbit (Figure 1) to form the first stepping stone in the assembly of the International Space Station. Much like the Hubble Space Telescope, the International Space Station has taken twenty years to advance from concept to reality. I remember the time scale well, because I began planning for utilization of the space station in 1984, during the phase A study. I also intend to see the full assembly through to completion — about five years from now.

G. Haskell and M. Rycroft (eds.), International Space Station, 185–194.
© 2000 *Kluwer Academic Publishers.*

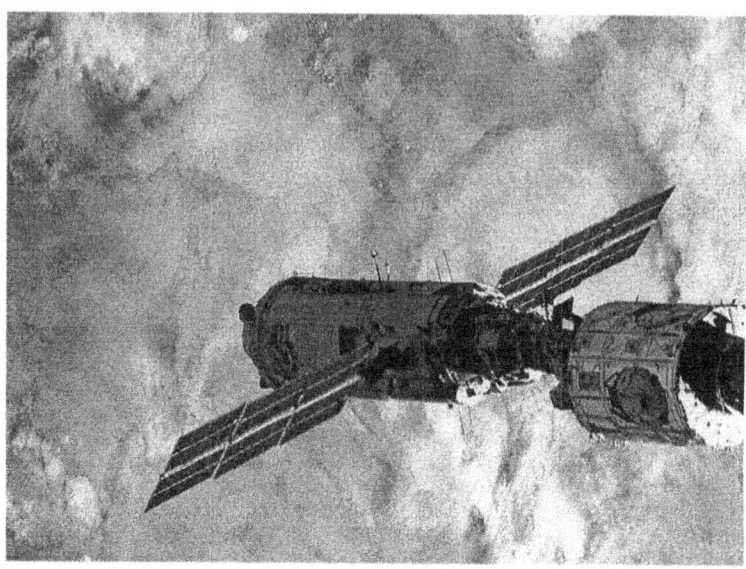

Figure 1: Zarya-Unity in low Earth orbit

Just as looking through the Hubble Space telescope has dramatically advanced our understanding of the cosmos, I am confident that the research to be performed in the space station laboratories, both inside and outside, will soon begin to expand our knowledge in basic biology, chemistry, physics, and engineering.

The reason why I am so confident is because an understanding of the forces that drive *motions* is a fundamental step in unveiling the secrets of dynamic systems — both organic and inorganic.

The Euler and Navier-Stokes equations form the basis for our current understanding of motions. Not only liquid and gaseous (fluid) systems, but also solids undergoing changes of state, are subject to the laws of motion. The force of gravity on unit mass, as readily apparent by the term "g", is pervasive in these analytical expressions (Figure 2). Whether it is the diffusion of biochemicals across a permeable membrane in living systems, or the dispersion of dopant atoms in the lattice structure of a solid state compound undergoing solidification, understanding the transfer of mass— *motion* — is the key to advancing our knowledge.

Fundamental Motion
(Euler and Navier Stokes Equations)

Local Inertia	Convective Inertia	Friction	Equations
Steady Motion or motion considered as a succession of steady motions	Slow Motion	w/o friction	$grad\ (p + \rho gz) = 0$
		w/ friction	$-grad\ (p + \rho gz) + \mu s^2 V = 0$
	Irrotational Motion	w/o friction	$grad\ (\rho\frac{V^2}{2} + p + \rho gz) = 0$
	Rotational Motion	w/ friction	$grad\ (\rho\frac{V^2}{2} + p + \rho gz) = -\rho(curl\ V) \times V + \mu s^2 V$
Unsteady Motion	Slow Motion	w/o friction	$\rho\ \frac{dV}{dt} + grad\ (p + \rho gz) = 0$
		w/ friction	$\rho\ \frac{dV}{dt} + grad(p + \rho gz) - \mu s^2 V = 0$
	Irrotational Motion	w/o friction	$\rho\ \frac{dV}{dt} + grad(\rho\frac{V^2}{2} + p + \rho gz) = 0$
	Rotational Motion	w/o friction	$grad\ (\rho\frac{V^2}{2} + p + \rho gz) + \rho\ \frac{dV}{dt} + \rho(curl\ V) \times V = 0$

Walter, H. U. (editor), Fluid Science and Materials Science in Space, Springer-Verlag, 1987

Figure 2: Euler and Navier-Stokes equations, with p pressure, ρ density, g acceleration due to gravity, z height, V velocity, and μ coefficient of viscosity

The microgravity environment in a space station orbiting the Earth represents a scientific frontier as exciting to researchers entering the 21st century as the vacuum environment was to investigators early in this century. Their work ultimately led from vacuum tubes to transistors and then to the large-scale, high-speed integrated circuits that drive contemporary technology. Studying the role of gravity in the fundamental equations of motion offers the prospect of not only improving our understanding of the forces of motion at the molecular level, but also, and more importantly, learning to control those forces.

Some have argued that the past fifteen years of Space Shuttle laboratory sorties to low Earth orbit have not yielded products of economic value. But closer study will reveal that during this period less than nine months of actual on-orbit research time have accrued. This, of course, leads me to one of the primary reasons why we are building a permanently crewed space station. It will operate continuously, for many years, with unprecedented laboratory capabilities.

Since the inception of the space station program, there has been an ongoing debate regarding the mission. Here, I suggest a resolution of that debate. The station has both intangible and tangible missions.

The intangible include global cooperation, the inspiration of our children to pursue excellence in science and technology, and, of course, the intrinsic human quest to explore. The more tangible missions encompass scientific research, technological advance, and economic development. It is this mission in economic development which leads to the subject of this paper.

2. NASA's Commercial Development Plan for the ISS

In November 1998, a significant convergence of events (see Figure 3) took place. The long awaited first element launch (FEL) milestone for the space station was finally achieved. The program entered its operational era— over the the next twenty years the largest civilian engineering project ever undertaken will play out in homes, classrooms, offices and industrial sites around the world.

November 1998 Convergence Of Events

1. 1998 Commercial Space Act (Public Law 105-303) passed by Congress and signed by President.

2. NASA "Commercial Development Plan for the International Space Station" approved and released.

3. International Space Station First Element Launch (FEL) milestone successfully achieved.

Figure 3: November 1998 Convergence of Events

The 105th Congress of the United States also passed, and the President signed into law, the Commercial Space Act of 1998. This legislation has created the impetus for a bold new initiative in the economic development of space, with the space station as a cornerstone. Also, in concert with the Commercial Space Act, NASA released a Commercial Development Plan for the International Space Station. This plan does not obviate the missions in scientific research or technological advance, but, for the first time, articulates a vision and a specific strategy for the third component — economic development. In all respects this mission — a mission for the private sector — follows logically and directly from the NASA missions in applied science and basic technology.

The NASA plan (Figure 4) has been throughly reviewed and formally approved by the Associate Adminstrators for Space Flight and for Life and Microgravity Sciences. It will shortly be submitted to the Congress by the NASA Administrator in response to specific provisions of the Commercial Space Act.

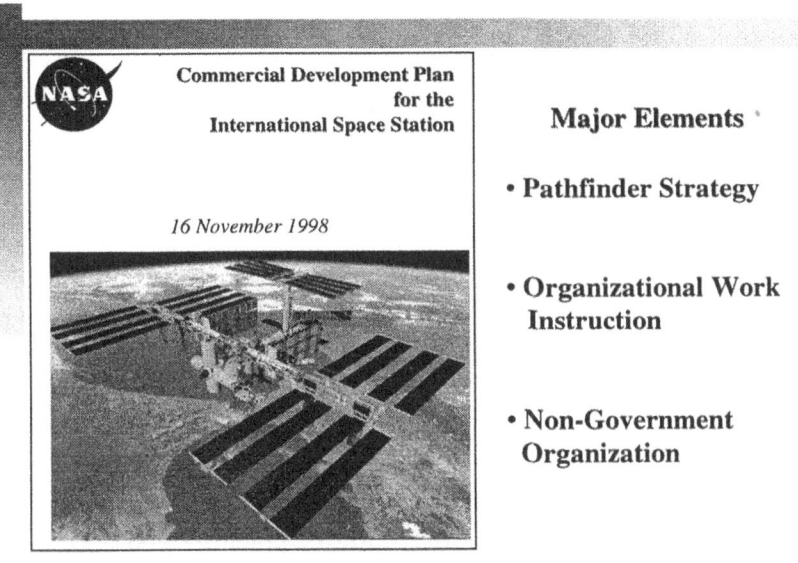

Figure 4: NASA's Commercial Development Plan for the ISS

There are three major elements of the Commercial Development Plan for the International Space Station; the plan is built upon a pathfinder strategy (Figure 5). The time has come to leave behind generalizations and hypothetical circumstances, and turn instead to specific business proposals. Over the years many barriers to private investment have been discussed, debated and dismissed, to no avail. The only pragmatic way to break down these barriers is to identify, in detail, specific privately sponsored enterprises – and then to use these enterprises as test cases to drive change in government policies, practices and procedures.

Pathfinder Strategy

• **Internal study identified potential commercial opportunities across ISS scope:**
   [3] **utilization**
   [3] **sustaining operations**
   [3] **evolutionary development**

• **External market assessment complete.**

• **External audit of costs complete.**

Pathfinders will be used to break down barriers and open the path for economic expansion.

Figure 5: NASA's strategy is to break down barriers to private investment and enterprises aboard the ISS

NASA has completed an internal study of potential pathfinder areas for commercial development. The study suggests that business opportunities may exist across the entire spectrum of the International Space Station program: utilization, operations and future developments. In the area of utilization, the station offers distinct comparative advantages for the development of advanced engineering technologies – components and subsystems can be tested and modified in real time without the expense, schedule, or risk of multiple launches. New sensors and detectors are ready examples.

In the area of operations, commercially provided services may represent cost savings, or cost avoidances, in comparison to government owned and operated systems. New transportation vehicles and carriers are ready examples. In the area of future capability development, growth elements of the station, perhaps co-orbiting, represent commercial opportunities. Private development of an International Space Station habitation module, with medical, recreation and health facilities could become the precursor to future complexes as envisioned by Arthur C. Clarke and cinematically brought to life over 30 years ago by the late Stanley Kubrick in "2001 — A Space Odyssey". With the space station becoming a reality, these concepts are no longer "far out".

NASA is now open to such entrepreneurial visions, and the space station is now open for business. NASA is accepting commercial offers with the intention thoroughly to evaluate all prospects and proceed with the most pragmatic pathfinders. But NASA cannot proceed alone; there must be clear industry sponsorship, with private investment. NASA will not be releasing requests for proposals (RFPs) nor awarding government funded contracts. There is no budget for this initiative, neither will NASA be requesting an appropriation. There are real opportunities for the private sector.

Next, in addition to our internal studies, NASA has initiated an independent market assessment as required by the Commercial Space Act; this study is due out this month. NASA has initiated a retrospective analysis of the long- and short-run marginal and average costs related to Space Shuttle operations, as well as a prospective assessment of the analogous cost projections for the International Space Station. This information will form the basis of estimates from which we will structure a pricing policy for the ISS.

The absence of a definitive pricing policy has long been viewed by the private sector as an obstacle to commercial development; NASA intends to remove that obstacle. NASA is working with the necessary stakeholders in the executive and legislative branches to converge on a value-based pricing policy with a marginal cost floor. NASA is also requesting the authority to waive all, or part, of the marginal costs in the short run, in order to stimulate industrial investment, while invoking the full marginal costs in the long run, in order to ensure that the government does not end up in the position, a decade from now, where profitable enterprises have been created which rely on public subsidies. NASA has asked industry to participate in identifying specific pathfinders.

Thus, NASA is moving rapidly toward a more substantive engagement with the private sector. NASA has already received seven commercial offers and projects another five in the next quarter. The prospect of incoming proposals brings us to the second major element of NASA's Commercial Development Plan, namely Organizational Work Instruction (Figure 6).

Organizational Work Instruction (OWI)

• **Commercial offer registration at NASA headquarters.**

• **To be developed under ISO 9000 standard.**

• **Will establish "single-point-of-entry".**

• **New NASA HQ "Space Utilization and Product Development Division" in place.**

> *The OWI will be used to streamline and discipline the processing of commercial offers.*

Figure 6: Organizational Work Instruction (OWI)

Historically, commercial concepts have entered NASA through a variety of doors, both at headquarters and in the field. The situation reminds me of first year University courses in Physics, and fundamental wave theory. When wave sources are not synchronized, destructive interference results. Synchronization of the wave sources is necessary in order to achieve constructive interference. This is where NASA is headed with the second element of the plan.

There has been a NASA-wide effort underway to obtain ISO-9000 certification. Under the auspices of this standard NASA is developing an Organizational Work Instruction (OWI) that will register all commercial offers asociated with the International Space Station. In the future all formal offers will enter through a single door. That is not to say that NASA will discourage informal discussions of commercial concepts across the agency, at all levels. This should, and must, continue. Rather, it is intended to bring order and discipline to the process, once the private organization has reached the stage at which it is prepared to submit a formal offer.

The Commercial Space Act requires NASA to report on International Space Station commercial proposals received in calendar years 1997 and 1998. As this report was assembled, it was extremely difficult to identify and track such proposals; that led to the conclusion that the process badly needed reform.

NASA's objective will be to establish a fully auditable process that treats proprietary proposals with a much higher level of confidentiality. The process will also highlight where the obstacles lie, in specific terms, and what NASA's rationale is for acceptance or rejection.

People often ask: "what are the selection criteria?" Quite simply, the ratio of private to public funding will be the principal figure-of-merit. In addition, the presence of non-government markets will be a significant factor. In time, NASA expects to develop performance metrics to track the time required to the disposition of commercial offerings, as well as the success rate in reaching formal agreements in various categories.

Now one might ask: "should the government even be in the position of evaluating and selecting commercial ventures? " The answer, of course, is no. This is the role of the capital markets and, in the United States, the market is particularly adept at this. This leads to the final element of NASA's plan — the concept of a non-government organization (NGO) to manage International Space Station utilization and economic development (Figure 7).

Non-Government Organization (NGO)

• **Management of Space Station Utilization and Economic Development.**

• **Reference model developed.**

• **National Research Council evaluation underway.**

• **Trade studies initiated on NGO forms:**
   [3] **direct contract**
   [3] **cooperative agreement**
   [3] **government corporation**

The NGO will be used to undertake those actions which are beyond the scope of the government sector.

Figure 7: NASA plans a Non-Government Organization to manage ISS utilization and commercial development

This is, perhaps, the most controversial element of the plan, despite the fact that NASA has a terrific history of success with NGOs. When NASA was created the Jet Propulsion Laboratory, a federally funded R&D center operated by the California Institute of Technology under a contract with NASA, became a closely associated element of the civil space program. A few years later the Communications Satellite Corporation — COMSAT — was created as a government corporation and as a signatory to the INTELSAT organization. More recently, the Space Telescope Science Institute began operating under a contract with the Association of Universities for Research in Astronomy (AURA). And now the National Space Biomedical Research Institute has (NSBRI) been established through a cooperative agreement under provisions of the Chiles Act.

Among the various legal options — a government corporation, a cooperative agreement, or a fixed price contract — it is not yet clear which, if any, would be suitable for a space station NGO. That is why NASA is spending this year thoroughly evaluating the alternatives. It will be a great challenge forming an association that is capable of managing all three missions (components) of the International Space Station's Commercial Development Plan, not to mention ensuring upward compatibility to a global scale of operations in conjunction with the international partners. Perhaps the objective is too bold; if so, NASA is prepared to limit the scope of the NGO to commercial development.

We have enlisted the services of the National Research Council to assist in this effort and to ensure that the outcomes are pragmatic.

3. Conclusion

It is said that "nature abhors a vacuum"; perhaps, by shifting our laboratories to the microgravity and ultravacuum of space, nature's secrets will be further revealed. In the course of doing so a surrounding infrastructure will emerge, and the space economy will expand. The International Space Station will form the nucleus for human activities in space for the foreseeable future. However, it will not remain so indefinitely. Inevitably, this *Nucleus Station* will become only a portal to the next human frontier in space.

Enabling Better Science: A Commercial Communications Payload for the International Space Station

D. Beering, Infinite Global Infrastructures, LLC, 618 Maplewood Drive, Wheaton, IL 60187-1400, USA

e-mail: drbeering@sprynet.com

Abstract

During the past five years, using NASA's Advanced Communications Technology Satellite (ACTS), a group of NASA and industry participants have performed a series of experiments focusing on the interoperability of TCP/IP, ATM, and higher layer protocols and applications. These experiments have yielded very exciting results, including pro-forma configurations in the following areas:

- TCP/IP data transfer over geostationary satellite delays at speeds exceeding 500 Megabits per second using standard network hardware, computers, and operating systems
- Video, audio, and telephony over satellite links using ATM to engineer links with a constant Quality of Service for these time-sensitive applications
- Security overlays featuring encryption and IP firewalls at up to 155 Megabits per second
- Mobile satellite terminals that operate on ships, trucks, aircraft, and (eventually) spacecraft.

This paper describes a proposed communications payload for the International Space Station, which supports the use of commodity industry-standard communications protocols to support direct user access to science instruments and experiment payloads from the ground. The payload concept, which is based entirely on commercial off-the-shelf products, was developed as a result of the five-year ACTS experiments program.

1. The ARIES Project

Sponsored by the American Petroleum Institute, the ATM Research & Industrial Enterprise Study, or ARIES, involved more than thirty US organizations ranging from NASA laboratories to major oil companies, to commercial communications providers and telecommunications equipment manufacturers. The goal of the project was to study the emerging class of high-performance virtual networks by building a pro-forma model of a service provider-based, high-performance network. The motivation behind ARIES was to position the oil industry to assimilate new communications technologies rapidly in order to enable the industry to collapse cycle times dramatically for US-based and remote operations.

G. Haskell and M. Rycroft (eds.), International Space Station, 195-202.
© 2000 *Kluwer Academic Publishers.*

Modern exploration operations generate significant volumes of complex data and telemetry that need to be transmitted in a timely fashion to decision makers at the home office. Since the petroleum industry operates in nearly every region of the world, the communications infrastructure must be global. Further, since wired, fiber infrastructure is not available in most frontiers where the industry operates, the communications infrastructure must be wireless or satellite-based. The most daunting of the remote data transfer challenges involves the transfer of seismic data from moving seismic acquisition vessels while they are on a prospective exploration site. This operation encompasses remote communications, high-speed data transfer, and mobility all in the same model. It was this challenge that ARIES pursued with the most vigor, for to solve this problem would mean that most of the other problems involving remote connectivity would have to be solved along the way.

Due to the high costs of developing fossil fuel resources in new frontiers, consortium-based (multiple oil companies) exploration has become the preferred way to open up these prospects. The ARIES experiments were conducted assuming consortium-based exploration as a baseline. From a networking perspective, successful consortium-based exploration leverages the following components:

- A shared high performance terrestrial network connecting all of the key participants from the consortium, or geographically separated participants from a single organization. This allows computation, modeling, interpretation, and visualization to be distributed geographically. If the network connecting the different locations is capable enough, the physical location of each resource becomes irrelevant
- A high-speed link from remote data acquisition resources. Since the remote acquisition resources move, this would take the form of a high performance satellite channel. The primary benefit of this activity would be to make unprocessed data available to the center of excellence in near real-time, reducing the need to place costly human resources at the acquisition sites (single-tasked). This also allows rapid decision making based on earlier access to the unprocessed data
- Intelligent use of data pre-processing and data compression at the remote site. This would reduce the reliance on the high-speed satellite channel, allowing it to be used as needed, rather than full-time.

The telecommunications infrastructure supporting the ARIES project was based entirely on industry standards, most notably TCP/IP and Asynchronous Transfer Mode (ATM). The experiments focused on the following elements:

- Experimental work with TCP/IP to enable the Internet protocol to scale to accommodate high-speed, reliable data transfer over satellite channels with very high bandwidth and geostationary satellite delays
- Development of applications to facilitate high speed reliable data transfer from high-performance, high-density tape subsystems, in addition to (much easier) host-to-host transfers
- Development of stabilized shipboard platforms to support accurate pointing and steering of Ka-band satellite antennas to counter ship motion at sea
- Development of end-to-end models of collaboration using very complex graphical data sets and high-performance terrestrial extensions of the ship-to-shore satellite link.

Early in 1996, a fully functional model of the interactive oil exploration concept was built by the ARIES team. The demonstration illuminated the areas where further work was necessary in order to make the vision of interactive exploration a reality. The experiment culminated in a demonstration of the system at the National Press Club in Washington, DC. During the demonstration, a high-speed shipboard data transfer was performed from the M/V Geco Diamond, a seismic acquisition vessel that was operating in the Gulf of Mexico, to two supercomputing facilities in the continental US. The satellite link for the demonstration operated at 2 Megabits per second over NASA's Advanced Communications Technology Satellite, operated by the NASA Glenn Research Center in Cleveland, OH.

Many important lessons were learned, including:

- Multi-service high-speed networks can be successfully deployed to moving ships using small, articulated (steered) antennas. In the case of the Diamond experiment, the Ka-band antenna measured 0.40m by 0.11m
- The best networking technology for supporting multi-mission remote networking is Asynchronous Transfer Mode. This technology provides the ability to add applications to the mission profile on very short notice without having to redesign the entire data transfer application. ATM also allows different applications to be supported with different qualities of service (dedicated, best effort, etc.)
- The best networking technology for reliable data transfer is TCP/IP. In a point-to-point data transfer application, TCP/IP guarantees delivery through the use of acknowledgements. Therefore, using TCP/IP over ATM provided a networking model that supported the greatest flexibility of data networking alternatives, while guaranteeing delivery of the most critical data

- The data transfer requirements for shipboard seismic data are not symmetrical. The ship-to-shore satellite channel needs to operate at a much higher data rate than the shore-to-ship channel. The shore-to-ship data rate must be sufficiently large to support the required TCP/IP acknowledgements, however. For TCP/IP-based data transfer, this data rate can be as little as one or two percent of the ship-to-shore channel data rate.

Many of the insights gleaned from the three-year ARIES project are applicable to the space communications application area.

2. The 118 Series of ACTS Experiments

The ultimate goal of all of the oil industry experiments was to use TCP/IP for the end-to-end communication between ships and remote sites, and the researchers. The reason for this was simple — applications based on available industry standards are substantially less expensive to develop, maintain, and operate than those based on proprietary protocols and gateways. This goal is coincident with NASA's Consolidated Space Operations Contract (CSOC). The network architecture for CSOC is referred to as the Integrated Operations Architecture, or IOA. During 1997, 1998, and 1999, NASA has performed a series of very high data rate experiments using the Advanced Communications Technology Satellite. The series of experiments, all which operated at 622 Megabits per second across the satellite link, were referred to as the *118* series of experiments.

Until recently, it was believed that special applications and protocols would need to be developed to facilitate moving data across satellite links at hundreds of Megabits per second, due to the inherent limitations of the original TCP/IP protocol suite. Recent enhancements to the TCP/IP protocol have made it possible to support very high data rates using off-the-shelf equipment, however. Many of the leading computer and operating systems manufacturers are now shipping their operating systems with these enhancements implemented, or at least resident. The enhancements are well documented in the Internet Engineering Task Force's TCP/IP recommendations, entitled RFC 1323 TCP Extended Windows and RFC 2018 TCP Selective Acknowledgement.

The most recent ACTS experiment, called 118Next, is using the Advanced Communications Technology Satellite to test computing hardware and operating system software from several of the leading computer and communications vendors to determine the level of interoperability among the vendors at high rates using TCP/IP. The performance of each individual vendor's TCP/IP implementation is also being studied.

Recent tests have yielded the following results over the satellite link:

* Workstation to workstation best sustained TCP/IP transfer rate of 520 Megabits per second
* Tape-to-tape best sustained transfer rate of 324 Megabits per second.

In October 1998, the same team of NASA and industry researchers combined their efforts with the Satellite and Wireless Networking Branch of the US Naval Research Laboratory to field a stabilized satellite terminal on a 14m vessel on Lake Michigan off the coast of Chicago, IL. The system utilized a one-meter antenna and pedestal manufactured by SeaTel, Inc.. For the duration of the experiment, the satellite link between the vessel and the Glenn Research Center operated at 45 Megabits per second. During the at-sea tests, ship-to-shore sustained data transfer rates of 40.5 Megabits per second were achieved using TCP/IP over ATM on the satellite link. The link also supported real-time full-motion video, CD-quality audio, and connections to the Internet /World Wide Web.

3. Application to Robotic (Science-Gathering) Space Vehicles

As a part of the development of the Integrated Operations Architecture (IOA) for the Consolidated Space Operations Contract, Lockheed Martin asserted that TCP/IP and ATM could be applied uniformly across NASA's space architecture for reliable, end-to-end communications at substantially lower cost than current systems. Pursuing that claim, Lockheed Martin built a pro-forma network configuration featuring a simulated science-gathering spacecraft communicating with the ground through a link that simulated NASA's Tracking & Data Relay Satellite. The simulated spacecraft was connected to a control center in Houston, TX, which was in turn connected to a large Internetwork. The experiment was publicly demonstrated in January 1998.

In actuality, the simulated spacecraft was located in building 55 at the NASA Glenn Research Center. The spacecraft was constructed using a series of software models running on a pair of computer workstations, including an IBM AIX host, and a Sun Solaris host. The simulated TDRSS link was a real satellite link, carried across NASA's Advanced Communications Technology Satellite. The satellite link was asymmetric, with the return link (spacecraft-to-ground) operating at 45 Megabits per second, and the forward link (ground-to-spacecraft) operating at 4 Megabits per second. The following applications were supported in the experiment:

- High-speed reliable data transfer using TCP/IP optimizations developed during the 118i and 118j experiments
- Interactive TCP/IP sessions between the spacecraft and the control center carrying telemetry
- Real-time, full-motion video (bi-directional)
- Real-time telephony (bi-directional)
- Multi-level security, including dual-key, triple DES encryption applied at the ATM layer, per application
- Extensive use of automation onboard the spacecraft and in the ground systems
- Enforcement of varying Qualities of Service (QoS) for different applications using the link.

The demonstration provided a great deal of encouragement that the vision of an end-to-end space communications architecture based entirely on recognized commodity commercial standards was achievable.

4. Application to Human-Rated Space Vehicles

The demonstration of real-time, bi-directional services from the simulated science spacecraft was designed to illustrate the CSOC IOA applying equally well to human-rated space vehicles as to robotic spacecraft. Following the successful demonstration, both Lockheed Martin and NASA Glenn started independent studies to determine the feasibility of placing a Blackbird-like payload in a human-rated space vehicle, starting with a series of experiments on the Space Shuttle, and moving later to the International Space Station. The two projects had the following attributes:

- Based entirely on commercial communications protocols (TCP/IP and ATM)
- Envisioned starting on the Space Shuttle, and later on the International Space Station
- Envisioned an experiment package supporting a host of recognized commercial interfaces — Ethernet, ATM, V.35, etc.
- Proposed to start with a Ka-band experiment on ACTS
- Realized that antennas supporting high data rates would need to be developed, or borrowed from the existing Shuttle inventory.

5. Conclusion

The end goal of this work is to utilize commercially available satellite services and ground distribution networks to provide the necessary space-to-ground data networking connectivity through the standards-based communications system on the ISS. However, it will be some time before viable commercial satellite platforms exist. Therefore, in the meantime, NASA will need to leverage existing space networking assets, in particular the Tracking & Data Relay Satellite System (TDRSS). While this goal can be achieved today utilizing NASA's existing space network infrastructure, the system will be more cost effective when the space network (space-to-ground link) is commercially provided.

The work discussed here involves the implementation of an onboard communications infrastructure that relies on commodity industry standard interfaces and protocols throughout (see Figure 1). This is the most expeditious way to prepare for utilization of commercial satellite/ground network assets for the future ISS communications system. The combination of the advanced onboard system with a high-performance standards-based ground distribution network will be an enabler for science data to be carried seamlessly between the vehicle and the scientists and operations personnel.

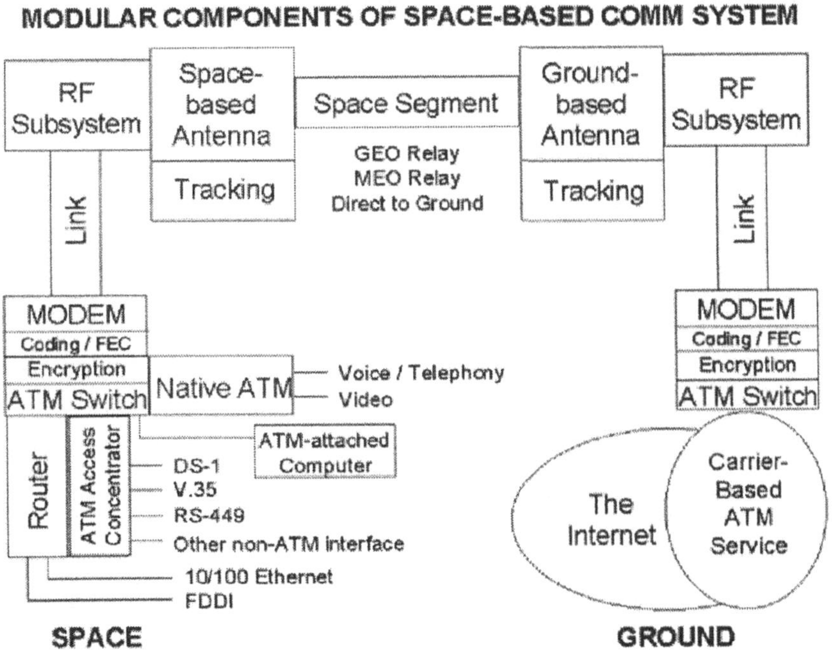

MODULAR COMPONENTS OF SPACE-BASED COMM SYSTEM

Figure 1. Block Diagram of the ISS Communications Architecture

The combined system (spacecraft, space network, and ground network) will be able to support nearly any application that can be realistically envisioned at nearly any location on the ground. The science community would benefit dramatically from such an investment. Without a state-of-the-art communications system, the ISS is just an outpost.

Commercialization of Management Know-How Generated by the ISS Program

M. Bosch, University of Regensburg, Faculty of Business Administration, D - 93040 Regensburg, Germany

e-mail: Michael.Bosch@wiwi.uni-regensburg.de

Abstract
The International Space Station (ISS) is the largest and most complex technology project in the history of humankind. Currently, 15 countries and their aerospace industrial contractors are participating in this program. This leads to the following organizational challenges for program management: (1) Interdisciplinary integration of highly specialized professionals, institutes and subcontractors, (2) International and intercultural integration of participating countries, (3) Life-cycle-oriented integration: experts for the later stages of space station utilization and operation are also included in the early phases of design and development, (4) Integration of different users and their experiments during the Space Station utilization phase. To face these challenges, NASA and the prime contractor Boeing established a team-oriented organization with Integrated Product Teams (IPTs) and Analysis and Integration Teams (AITs). This type of organization successfully meets the integration requirements mentioned above.
Similar integration problems occur in other projects both within as well as outside the aerospace sector. Even in the real estate development business, there is a growing awareness that a life-cycle-oriented approach to project management is crucial for project success. Such an approach integrates the utilization phase in an overall facility management. The goal of this paper is to show ways to commercialize this management know-how generated by the ISS-Program effectively. This paper begins with an explanation of the IPT/AIT-organizational structure in the ISS-Program, and then explores possibilities to create commercial consulting services for the distribution of this management know-how. Finally, the successful implementation of an ISS-inspired, team-oriented organization in a large real estate development enterprise is described.

1. Introduction

The development and production of spaceflight systems present challenges which can hardly be compared to other branches of engineering. Especially for manned missions, technically perfect system solutions need to be developed in order to assure safe missions.

Even for missions carried out by a single country, an organization which guarantees cooperation between main contractors and subcontractors, government agencies, universities and research institutes is necessary. At the current time, 15 nations are involved in the ISS Program. This increases the level of complexity in comparison to projects carried out by a single country.

G. Haskell and M. Rycroft (eds.), International Space Station, 203-209.
© 2000 *Kluwer Academic Publishers.*

2. Integration Requirements

2.1 Interdisciplinary Integration

A project as large and as complex as the ISS requires a division of labor among highly qualified specialists, departments, enterprises and institutions. In addition to technicians, experts in the natural sciences, computer sciences, medicine, business, law and for the operation and utilization of orbital facilities are also necessary. The project organization must guarantee an interdisciplinary integration of specialists and contractors.

2.2 International Integration

The international cooperation between the 15 countries participating in the ISS Program requires additional management efforts because:

- Development and finances must be regulated by contracts between the participating countries
- Different languages, cultures and legal systems must be taken into account
- Of the ever-present risk that one of the international partners may withdraw from the project; such situations must be covered by contingency plans.

A major problem in the management of international, government-funded space programs carried out by the European Space Agency (ESA) is "juste retour", the "geographical return rule". This means that, for a specific program, the contract volume awarded to the industrial contractors of each Member State must be nearly equal to that state's payment to ESA for this program.

2.3 Life-cycle-oriented Integration

Each project goes through a certain life-cycle, which can be divided into the following phases:

- Concept Phase with Feasibility Studies (Phase A)
- Definition and Design (Phase B)
- Development (Phase C)
- Manufacturing (Phase D)
- Operations and Utilization (Phase E)
- Dismantling and Disposal (Phase F).

A large portion of the costs for the entire life-cycle are already determined at the end of the Definition and Design Phase [Reference 1]. As a result, optimization of life-cycle costs requires the early integration (in Phases A and B) of experts assigned to later project phases.

3. Integrated Product Team Organization

To fulfill these integration requirements, NASA and the prime contractor Boeing established a team-oriented organization with Integrated Product Teams (IPTs) and Analysis and Integration Teams (AITs). In this type of organization, an interdisciplinary group of people is responsible for the design, development, manufacturing, operations and support of a specific "product" [Reference 2]. A product is defined as either a hardware or software element of the ISS, a document, a procedure, a plan or a facility. Representatives of the following disciplines may be included in an IPT, as needed:

- Program Control
- Systems Engineering
- Design
- Development
- Manufacturing
- Quality Assurance
- System Safety
- Test
- Operations
- Utilization
- Crew.

In addition to employees of the contractor, an IPT can also be made up of additional personnel from NASA (e.g. Space Shuttle Program), as well as from other contractors or international partners. An IPT has all the resources required for its product and is accountable for technical, schedule and cost performance [Reference 3].

AITs provide system level analysis and integration for the IPTs. They ensure interface, configuration and integration control. Each contractor's *Program Manager* is responsible for the development of the work packages as specified in the contract. The program manager has full authority over all IPT's and AIT's within the enterprise. In addition, the program manager is also responsible for the project functionals [Reference 4]. The integration requirements mentioned above are successfully met by the organizational concept shown in Fig. 1, as introduced in 1994.

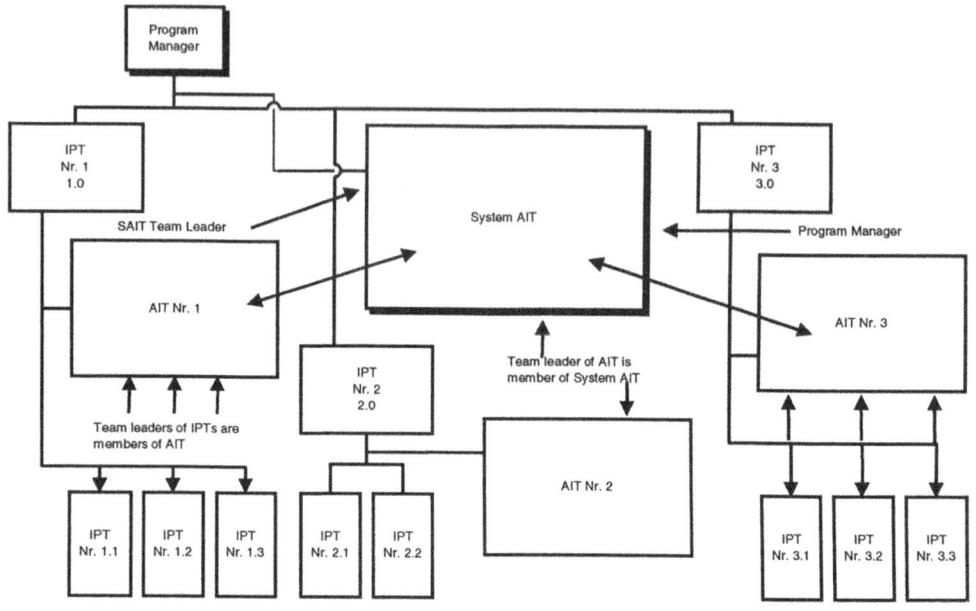

Figure 1. IPT-/AIT-Structure [Reference 4]

4. Opportunities for Commercialization

Similar integration problems occur in other projects both within as well as outside of the aerospace sector. In this section, possibilities to transfer organizational concepts from the ISS to other consulting and organizational projects will be explored. Team-oriented organizational structures can be applied to projects which have integration problems similar to those which occur in the Space Station Program. Projects can be divided into different classes, according to how well the organizational structures from the Space Station Program can be applied.

National and international aerospace programs have the most complex life-cycles and extended operation and utilization phases. Thus, they are an ideal application for team-oriented organizational structures. Large projects in non-aerospace branches must possess the following characteristics in order for an application of ISS organizational concepts to be useful [Reference 5]:

- high level of complexity with a large number of project participants
- high level of innovation
- high technical risks
- hard deadlines
- high risk of exceeding planned costs and deadlines.

Examples of suitable projects would be the development and operation of new transport systems, such as Mag-Lev trains, complex military systems (missiles, aircraft carriers) or the construction and operation of large, complex facilities, such as airports or power plants.

Smaller, less complex projects with fewer participants would not be suitable for a complete transfer of the ISS organization. The application of a few selected concepts, however, such as life-cycle oriented organization and integration of teams, could be of value.

Each case must be examined individually. A simple one-to-one transfer should be avoided. This new market niche for qualified analytical services can best be filled by professional management consultants.

5. Application in Real Estate Development

Real estate projects have life-cycles which include the following phases: conception, planning, construction and sale. In addition, they also have relatively long operation and utilization phases. The costs during the operation and utilization phases are much higher than those in the earlier phases. In order to plan an optimization of the entire project life-cycle, experts for the operations and utilization phases need to be integrated in the earlier project phases. A team-oriented, organizational concept for a large real estate developer, which includes the development, operation and utilization of a building is depicted in Fig. 2.

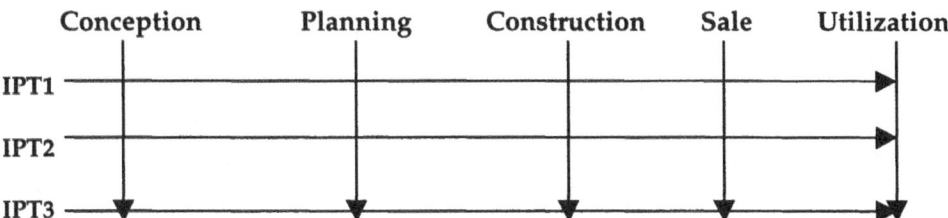

Figure 2. Life-cycle-oriented real estate development

An IPT is responsible for a given real estate project from its conception through its utilization. The team can consist of representatives of the following disciplines, as needed:

- Project Acquisition
- Architecture
- Civil Engineering
- Project Control
- Real Estate Sales
- Financing
- Property management.

Personnel from subcontractors or from the customer can be included in the individual teams, as needed. In this manner, the design of the building can be taken into account during the planning stage in order to simplify cleaning during the later operation and utilization phases. Consideration of different alternatives to reduce energy and water costs during the planning phase can thus result in a significant reduction of operation costs.

Previously, the life-cycle in the real estate development business has been largely defined to extend only up until the sale of the facility. The operations and utilization phases have been almost completely ignored. Both rapidly rising operational and maintenance costs as well as increasingly cost-critical customers require a radical change of course. In addition to the cost advantages already mentioned, a real estate developer or one of its subsidiaries can take advantage of the rapidly growing market for professional property management. Existing activities in the development, construction and sale of real estate bring advantages over other competitors in the acquisition of lucrative property management contracts.

6. Conclusions

The management concepts applied in the ISS Program offer many commercialization possibilities for professional management consultants. Transfer of the IPT organizational concept to other branches requires an individual consideration of the size and complexity of the project in question. Enterprises which previously defined their life-cycles to end with the development or production can acquire lucrative follow-up contracts, especially in the operations and utilization phases.

Acknowledgements
I would like to give special thanks to Patricia Shiroma Brockmann for the translation of this paper into English as well as for her continued encouragement and thought provoking discussions.

References
1. Krummrey, C., Blank, F.: Ermittlung von Lebenszykluskosten in der Raumfahrt, MBB-Publikation UR-E-910/86 PUB, 1986
2. National Aeronautics and Space Administration: International Space Station Alpha, Management and Implementation of Integrated Product Teams, Orientation Package, April 1994
3. Kennedy Space Center: Space Station AIT/IPT Overview, 1994
4. National Aeronautics and Space Administration: International Space Station Alpha, Management and Implementation of Integrated Product Teams, Orientation Package, April 1994
5. Bosch, M.: Management internationaler Raumfahrtprojekte, Gabler Verlag, Wiesbaden, 1997

An Initial Strategy for Commercial Industry Awareness of the International Space Station

C. Jorgensen, FDC, Inc., NASA Langley Research Center, 8 Langley Road, M.S. 328, Hampton, VA, 23681, USA

e-mail: c.jorgensen@larc.nasa.gov

Abstract
The on-orbit assembly of the International Space Station (ISS) began in December 1998. While plans are being developed to utilize the ISS for scientific research, and human and microgravity experiments, it is time to consider the future of the ISS as a world-wide commercial marketplace developed from a government owned, operated and controlled facility. Commercial industry will be able to seize this opportunity to utilize the ISS as a unique manufacturing platform and engineering testbed for advanced technology. Activities to allow the initial planning of the commercialization of the ISS have begun. NASA is currently in the strategic planning phase of the evolution and commercialization of the ISS, an essential and critical step. The Pre-Planned Program Improvement (P^3I) Working Group at NASA is assessing the future ISS needs and technology plans to enhance ISS performance. Plans are being formulated for ISS enhancements to accommodate commercial applications and the Human Exploration and Development of Space mission support. As this information develops, it is essential to disseminate this information to commercial industry, targeting not only the private and public space sector but also the non-aerospace commercial industries. An approach is presented for early dissemination of this information that includes ISS baseline system information, baseline utilization and operations plans, advanced technologies, future utilization opportunities, ISS evolution and Design Reference Missions (DRM). This information is being consolidated into the ISS Evolution Data Book, an initial source and tool to be used as catalyst in the commercial world for the generation of ideas and options to enhance the current capabilities of the ISS.

1. Introduction

The assembly of the International Space Station (ISS) is underway and within a relatively short period of time, by 2004, will be completed and fully operational. The NASA Administrator, Mr. Daniel S. Goldin, declared that 30% of the U.S. Laboratory space would be commercialized as part of the directive of the Commercial Space Act of 1998 [Reference 1]. Recently, Mr. Goldin stated that "nothing would please me more than if commercial demand for Station accommodations reached 40, 50 or even 80 percent" [Reference 2]. Any amount of commercial use of the ISS could distribute the resource burden of continual ISS operations between NASA and the commercial sector, which would be a plus for both sides. This vision of commercial use of the ISS up to 80% is a signal for all sides involved to act now in planning for commercial ventures on the ISS.

Commercialization is a "...private sector, profit-seeking entity using its own or borrowed and/or invested funds to carry out activities intended sooner or

G. Haskell and M. Rycroft (eds.), International Space Station, 211-218.
© 2000 *Kluwer Academic Publishers.*

later to result in products or services that can be sold at a profit through a market, either to government or non-government customers or to a mixture of the two" [Reference 3]. The current NASA thinking is that industry will utilize the ISS for commercial research to develop new technologies that can be used in terrestrial-based commodities. This can be taken a step further to use the ISS as a possible and probable production facility and as a facility in which products and services can be bought and sold (i.e. products, entertainment) and, hence, a space marketplace.

NASA has developed a plan for commercializing the ISS; however, there is a need for supplemental activities to occur now in addition to those included in NASA's plan. Acting now can ensure the timely and effective development of the ISS as a viable and acceptable marketplace by commercial industry.

2. Parallel Perspectives

2.1 Existing Strategy

The current NASA strategy for commercializing the ISS is outlined in the Commercial Development Plan for the International Space Station [Reference 4]. The tactics included in this strategy include: 1) an independent market assessment, 2) identification of barriers to market entry, and 3) the establishment of a non-government organization (NGO) for ISS utilization development. The time frame for the completion of the first two activities would be June 1999 and the end of 1999 for the establishment of an NGO. This strategy is well thought out with the necessary steps for creating a new marketplace; however, the focus here is primarily from the NASA perspective. The studies and analyses outlined in the plan are directed towards the NASA implementing commercialization of the ISS.

2.2 A Commercial Industry Perspective

Developing the ISS into a marketplace should also be addressed from the commercial industry perspective. The initial transition of this facility to a commercial marketplace could place a heavy financial burden on commercial firms. It is evident that in the U.S., at least, the government must develop cost cutting processes for ISS access to entice the commercial firms to utilize the facility. The government may initially have to subsidize some of the expenditures of commercial companies who desire to venture into this new marketplace. This transition phase will allow NASA, in conjunction with commercial industry, to increase the efficiency of ISS operations processes and thus reduce operating costs. These cost reductions will provide the commercial firms with higher returns on investments (ROI) and ultimately entice more commercial firms to enter the ISS space marketplace. In addition to the investment burden which commercial companies will face, the acquisition of programmatic and technical information pertaining to the feasibility of proposed ISS-based commercial ventures presents additional roadblocks.

Commercial industry must be made aware of the opportunities that are available on the ISS. This should happen now so that commercial industry can initiate their planning strategies as NASA has done. It is going to take time for commercial industry to accept the ISS as a viable marketplace. Industry executives and analysts will need to look at cost versus ROI. Industry will also need facility and utilization information to determine from an operational standpoint whether they can use the ISS as a marketplace. This information must be disseminated to industry quickly and thoroughly in order for informed decisions to be made. Currently, there are Commercial Space Centers (CSC), usually universities, working in conjunction with NASA as an avenue for industrial firms to approach the space program. When an industrial interest becomes convinced that a space research activity has a potential economic benefit, it can approach a CSC with a proposal [Reference 5]. From these initial discussions, the CSC will determine whether the space research activity is feasible and will work with them to integrate the research into the ISS Program.

The CSCs work with approximately 135 industrial firms at this time. This avenue has been an effective means for commercial industry to take to get to space. However, this has been primarily for research purposes only and many of the companies associated with the CSCs are already involved in the space industry in some manner. There are many firms who are not affiliated with any space related-industry who may be interested in utilizing the ISS as a marketplace in areas other than research. Additionally, there are many firms who have not even considered the possibility of utilizing the ISS in any manner for the mere fact that they are unaware that this is possible, or will be possible in the relatively near future. It is this part of commercial industry that this strategy targets, as is explained in the next section.

2.3 Commercial Industry Awareness

Many firms within the U.S. and worldwide are unaware of the fact that they could someday utilize the ISS for new product development in such areas as telecommunications, pharmaceuticals and materials processing, as a production facility, or even as a service provider, as in the entertainment industry. Even if they are aware of this fact, their planning on ways to utilize the ISS platform should begin now. For any industry to begin planning, they must have some initial source of information to help them determine potential venture characteristics such as cost, operations, facilities, resources, legal, potential for expansion, etc.. There is currently an overwhelming amount of information on the ISS that includes the planning and process documentation, technical specifications, assembly information and the list goes on. Not only is the amount of information mountainous, it is also mostly unavailable to the general public at this time. While this is understandable from the perspective of NASA and its contractors, from an industry perspective the unavailability of this information can be construed as another obstruction in their journey towards acceptance and utilization of the ISS. A commercial firm interested in obtaining information to establish the initial viability of a commercialization concept could find it to be an almost impossible task. Where would they begin? They would first have to determine what information they needed and request it from the owning entity, if they knew who that was. Even if they did get a positive response from the information owner, they would then have to sift through an enormous amount to extract the information which they need for their application.

Currently, an informational reference source is being developed at NASA Langley Research Center in Hampton, Virginia, sponsored by the Director of Advance Projects, Office of Space Flight at NASA Headquarters in Washington, D.C., as an initial solution to this potential roadblock. The document entitled "The International Space Station Evolution Data Book" [Reference 6] provides a focused look at the opportunities and drivers for the enhancement and evolution of the ISS during its assembly and beyond the assembly complete stage. These enhancements would expand and improve the current baseline capabilities of the ISS and help to facilitate the conversion of the ISS into a marketplace by and for the public sector.

The purpose of the data book is threefold. First, it provides a broad, integrated systems view of the current baseline design of the ISS systems and identifies potential growth and limitations of these systems. Secondly, it presents current and future options for the application of advanced technologies to these systems and discusses the impacts these enhancements may have on interrelated systems. Finally, it provides this information in a consolidated format to scientific and commercial entities to help generate ideas and options for creating new technologies and products, and to assist in determining potential beneficial uses of the ISS in commercial business.

3. ISS Evolution Data Book Description

The ISS Evolution Data Book is composed of six sections, the first of which is the introduction. The second, third and fourth sections give a broad, integrated systems view of the ISS baseline design. Section 2 of the data book provides a brief overview of each of the 22 major components of the ISS. This includes individual laboratories owned by the ISS International Partners, the integrated truss segments, the various nodes, propulsion modules, and science facility modules. These brief descriptions provide the readers with information on the overall Space Station.

Section 3 covers nine of the individual critical sub-systems onboard the ISS. These include Power; Thermal; Communications; Command and Data Handling; Guidance, Navigation and Control; Propulsion; Environmental Control and Life Support Systems; Robotics; and Structures and Mechanisms. The document provides high-level technical overviews of each system, the capabilities of each, the potential limitations of each, and a description of potential growth opportunities and limitations.

Section 4 provides information on the current baseline plans for the operation and utilization of the ISS. This section includes plans for utilization of crew time, traffic models, space availability, and resource availability including bandwidth and data rates for communications, power, thermal, and system ground commanding availability. It also includes descriptions of each of the internal facilities on board the ISS, which will provide specific services to payloads that are either currently planned, research-based activities or future commercial-type endeavors. These include facilities such as the Combustion Facility and Fluid Physics Facility that are used in materials research, the Life and Microgravity Science Gloveboxes which are fully contained facilities for performing biological or materials activities, and many more. Along the same lines, it provides location and resource information on the external payload facilities, including the U.S. and Japanese facilities. The European Space Agency facility and Russian facility will be described as information becomes available.

This baseline information will assist commercial industry in developing their view of the ISS as a potential marketplace. Industries can utilize this information to determine if their specific technologies can benefit the ISS and/or if the ISS offers an environment which they could utilize in a commercially viable manner.

The next two sections of the data book present current and future options for the application of advanced technologies. Section 5 presents the advanced technologies that are being investigated by the Pre-Planned Program Improvement (P^3I) Working Group, led by the NASA Johnson Space Center in Houston, Texas. These advanced technologies will provide enhanced capabilities to the ISS that may be beneficial to commercial industry. The section discusses proposed ISS technology enhancements that are known at this time and provides roadmaps for the investigation of each area. This information may entice commercial firms to look at their own technologies for enhancement of ISS systems.

Finally, Section 6 summarizes current Design Reference Missions (DRM) that are being investigated for post-assembly complete utilization and enhancements. These include free-flying satellite servicing, enhanced communications capabilities utilizing the Advanced Communications Tower and utilizing a new module, TransHab, for increased pressurized volume onboard the ISS (see Fig. 1). Each of these DRMs and the others that are presented in this document provide options for ISS enhancements that could be achieved via commercialization.

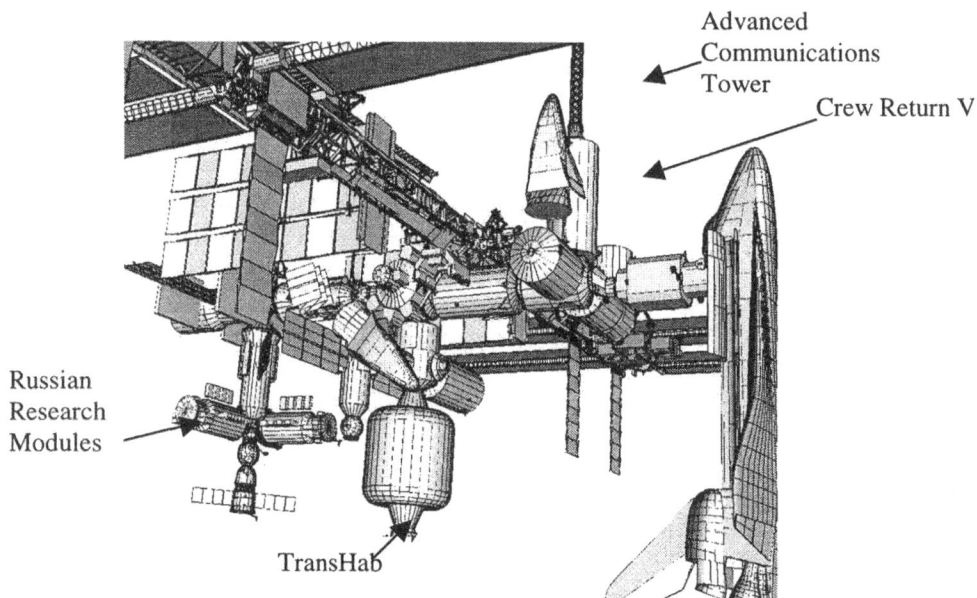

Figure 1. Concepts of ISS Evolution

4. Effective Utilization of the Data Book

The ISS Evolution Data Book is intended to provide high level technical information to the science community, commercial industry, academia, and the general public. Used as a desktop reference, the data book not only provides high-level finger-tip information on the technical aspects of the ISS, but also provides further references for more detailed information.

This data book should be used as a planning tool and a desk reference and not as a design tool for the development of a system, product, or any other entity. It is to be used for reference only to provide a high-level technical overview of the ISS, its capabilities, future options for enhancements, and opportunities for commercial use. It is hoped that the data book can act as a catalyst to facilitate innovative uses of the ISS for commercial ventures and to facilitate the application of non-aerospace technologies to enhance the many capabilities of the ISS.

5. Marketplace Outreach

The data book will be used as part of the marketing strategy for developing the ISS as an international marketplace. NASA Langley Research Center, in conjunction with NASA Headquarters, plans to distribute this information through the CSCs, through face-to-face contact at commercial and industrial trade shows, through the NASA commercialization initiatives at Headquarters, Johnson Space Center, and Langley Research Center, and via a publicly available web site. This web site is planned to have links not only to NASA public web sites, but also to business associations. As information becomes available, the document will be revised to remain current for industry. The updated information will be available electronically and via hard copy if desired. A database of interested firms will be established to ensure that the flow of information continues.

6. Summary

The commercialization of the ISS can be facilitated by informing industry of its potential and limitations. It is important that this information be disseminated now to as much of commercial industry as possible, both in the U.S. and globally. For the ISS to become a global marketplace, ISS-specific information must be communicated to guide the initiative and ambitions of public sector industry towards space. The use of the ISS Evolution Data Book to stimulate the creative expertise of industry is just one step towards commercializing the ISS. As the developing information is disseminated throughout industry, the acceptance and — hopefully — adoption of the ISS as a global marketplace will become a reality. However, this must be done now. We cannot wait until the ISS is fully complete and then hope that industry will fall in line. Providing the information to as widespread an audience as possible will help the commercial community to begin to look at the ISS as a marketplace now, and allow them to develop potential commercial uses for the ISS.

Acknowledgements
This work was supported by the NASA Langley Research Center in Hampton, Virginia, under contract NAS1-96013, with Mr. Jeff Antol as technical monitor. I would like to acknowledge the Spacecraft and Sensors Branch at NASA Langley Research Center.

References

1. Public Law 105-303 : Commercial Space Act of 1998, Section 101 from the 105th Congress, October 28, 1998
2. Goldin, Daniel S.: The National Importance of the Development of Space, Speech presented to the U.S. Chamber of Commerce at the Forum on the Future Development of Space, March 16, 1999
3. Logsdon, J.M.: Commercializing the International Space Station: current US thinking, *Space Policy, Vol. 14*, pp. 239-246, November 1998
4. NASA: Commercial Development Plan for the International Space Station, November 16, 1998
5. NASA: The NASA Research Plan, p. 41
6. NASA: International Space Station Evolution Data Book, Document No. Pending, Draft

Market Potential for the International Space Station (ISS) Service Sector

R. Nakagawa, National Aeronautics and Space Administration, Lyndon B. Johnson Space Center, Houston, Texas, USA

e-mail: roy.nakagawa@nasda.go.jp

R. Askew, National Aeronautics and Space Administration, Headquarters, Washington, D. C., USA

e-mail: raskew@mail.hq.nasa.gov

Abstract
There are two fronts to the commercialization of the International Space Station (ISS). One involves commercial use of ISS, while the other involves commercial service to ISS. This paper addresses the latter. How big is the market? Last year, United Space Alliance, the joint venture formed only a few years ago, came in second in a top 100 list of federal prime contractors. Their contract covers about a third of NASA's $3.2 billion annual budget for Space Shuttle operations. As other responsibilities are transferred to private industry, that percentage could double to two-thirds of NASA's shuttle budget. NASA's ISS budget may not be as big, but if one includes the projected operational costs of all of the international partners combined, the figure is quite high. What services could private industry provide?

This paper touches upon some of the possibilities, and also introduces how firms in Japan are well-positioned to capture pieces of the emerging market. The key to success will be the ability of the service provider to save costs, reduce risk, add value, and spin-off newly acquired know how. Examples include the use of commercial satellite communications service for the downlink of payload data, microgravity environment upgrades for improving science performance, telescience systems and robotic servicers to reduce crew workload and risk. Another more general way is for a service provider to facilitate the overcoming of traditional barriers and disincentives for commercial use of ISS.

1. Introduction to the ISS Service Sector

"We want to do everything possible, but what we do not want to do is pretend that it is privatization by having companies do 100% of their business with NASA and calling it commercial." Dan Goldin, during the House Appropriations Subcommittee on VA/HUD/IA Hearing on FY00 NASA Request, 23 March 1999.

The ISS is now a reality. The next generation full time laboratory in space can now be utilized in a limited way for scientific, engineering, and commercial initiatives. Much has been said about the commercial potential of this platform. Each of the international partners has promoted the ISS to industry to stimulate interest in using it to develop commercial data and possibly commercial

G. Haskell and M. Rycroft (eds.), International Space Station, 219-226.
© 2000 *Kluwer Academic Publishers.*

products. The amount of identified interest has been very limited. The primary reasons for this limited interest are:

- Industry does not see a clear reason for conducting activities in space (shortened product development time, significantly enhanced product/service, lower costs/higher margins)
- Commercial development issues, objectives, and goals are generally targeted to produce measurable progress within a year or less. The time to get a project into orbit has been, and is projected to continue to be, much longer than that required by private enterprise
- The reliability of access to space has been seen by the private sector as uncertain. With regard to NASA, industry sees the focus of the agency to be on the agency, not on the user
- Projected costs for access to space, even for activities requiring modest amounts of hardware mass to orbit and very limited times of operation, are well beyond the level of high risk investment which is acceptable.

Fundamental research in space and the development and qualifying of hardware to operate in space is a quite different area. Both offer significant opportunities for commercial activities on the ISS but will require significant changes in policies by each of the international partners. Historically the responsible government organizations have maintained tight control of the resources and their utilization. These agencies have made very few policy changes and implementations, which the private sector views as significant. The agencies believe in and operate space resources as their own; that is the way that these agencies evolved. In general, their personnel have invested careers in government service. There are few incentives to evolve from a controling to a supporting role.

Nonetheless, as the ISS platform evolves, the international partners will be operating the largest, most advanced, full time laboratory facility ever in space. During the past decade, the governments of the international partners have moved to transfer some government operated terrestrial facilities to operation by the private sector. Such transfers have not been without their problems, yet many have occurred. The results have, in general, been good. New commercial support services have arisen and many have expanded their role to include similar or spin-off services to the broader economic market. This is an area in which the ISS has a very significant commercial potential, but to achieve it will require fundamental changes in philosophy by each international partner.

2. Projected Size of the ISS Service Sector

NASA has developed ISS operational cost models and the partners have all accepted target figures, which exceed US $ 1 billion annually. In addition, to develop and conduct activities on the ISS, fully utilizing the resources, will require a significant annual expenditure. The planned research budgets for these activities by the individual partners have been projected and are constantly being reviewed. However, there is little doubt that full utilization for research will require an expenditure in excess of US $ 300 million annually. These levels of expenditure provide a significant incentive for the private sector to develop the services needed by both the ISS and the users of the ISS commercially. The government agencies representing the partners must make it clear through policy changes and early initiatives that the agencies will no longer provide these services.

Each partner agency is concerned about both operations costs and the amount of research to be accomplished. These two budgets are clearly connected. The agencies are thus looking for assistance from the private sector to reduce operations, thus permitting greater funding for research. At the same time, industry must see economic opportunities for itself, and must see a reduction of government intervention and a high degree of government consistency.

There currently exists a number of large aerospace corporations which have significant space operations experience derived by supporting the various international partner space agencies and supporting the space communications activities, both public and private. Some have formed partnerships (e.g., US Alliance) to leverage their individual abilities. Others see niche opportunities, but only if they are not competing with partner space agencies.

3. Some Examples of ISS Service

The ISS partners must operate the ISS as a laboratory for the conduct of research. The elements of operation are clearly known. For the private sector to assume specific roles, it must see clear evidence that the partner agencies will not compete with them. They must find ways by which to provide these services without increasing the costs to the partners and, at the same time, expect a reasonable rate of return. To do so will require innovation. Many innovative concepts have previously been proposed but few are being developed because the partners have not clearly committed to using private services if available. Some examples of potential areas of operations enhancements are:

- The single most significant element of cost is transportation. While work on the next generation launch vehicle continues at NASA, other unmanned vehicles are being developed by the partners. How such systems operate will determine the level of cost reduction
- Partner agencies currently maintain and operate their own communications infrastructure. Use of a satellite communications network would permit the reduction of partner-sustained facilities
- Telescience systems could increase the amount of research to be done by reducing the use of crew time for basic operations
- Improved thermal management capabilities within rack payloads and improved vibration isolation for long duration experiments would increase the research output.

4. Current Situation in Japan

The Japanese contribution to the ISS Program consists mainly of the Japanese Experiment Module (JEM), the H-II Transfer Vehicle (HTV), and the Centrifuge Module which the Japanese are developing as a cost offset to JEM launch on the Space Shuttle. Although Japan, as in Europe and the U.S., has long recognized the need to promote private industry use of the ISS, many barriers common to their partners as well as unique cultural barriers appear to exist. The authors suggest that these barriers, like those of the partners, are not insurmountable; to overcome them requires acknowledgment of their existence and a commitment to surmount them.

4.1 Overview of Japan's Space Industry

The Japanese space industry is supported primarily by government sales; therefore, a private sector-driven market has not yet arisen. In 1993, sales to NASDA accounted for 43% of the industry's total sales, while sales to other government agencies accounted for an additional 40% [Reference 1]. Since 1993, NASDA's annual budget has been rather flat, going from US $ 1.3 billion to the current level of US $ 1.6 billion. Meanwhile, the percentage outlay for space utilization promotion activities (which includes Japan's ISS contributions) has varied from a high of 33% in 1994 to a low of 22% in 1998 [Reference 2]. If a flat budget and an average percentage of 20% is assumed for the near future, the size of the government sector of the ISS market in Japan is roughly US $ 320 million per year, a significant financial incentive.

There are some signs that the Japanese space industry will grow more rapidly in the new millennium. The Ministry of International Trade and Industry (MITI), which is charged with promoting the growth of domestic industries, predicts that the market will triple by 2010, especially in the multimedia telecommunications field. The same report points out that, for Japanese companies to meet the coming global challenges, they must forge an integrated systems approach rather than simply remain as hardware providers [Reference 3].

4.2 Background on Promotion of JEM Utilization

Recognizing the limitations in government funds available to finance the utilization of the JEM, MITI along with the Science and Technology Agency (STA) enlisted the cooperation of a broad range of private firms to promote the use of the ISS by the private sector by forming the Japan Space Utilization Promotion Center (JSUP), in 1986. Since then, JSUP has published numerous reports (including a monthly newsletter), organized countless symposia and workshops, conducted seminars, and coordinated experiments on the shuttle as well as sounding rockets and drop towers.

Between 1993 and 1998, JSUP organized teams of researchers from national research institutes, universities, and corporations as members of the Space Utilization Frontier Joint Research projects, which were intended to serve as precursors to JEM utilization [Reference 4]. Although significant results were obtained from these projects in each of the areas of microgravity utilization, human space technology, and engineering research, industry continues to view the JEM as an expensive science laboratory with minimal payoff [Reference 5].

In an effort to make JEM utilization more attractive to industry, this year JSUP launched the Applied Research Pilot Project for the Industrial Use of Space (ARPPIUS) [Reference 6]. This pilot project is similar to the Microgravity Applications Promotions (MAP) Program at ESA. Both programs recognize the need to shorten turnaround time, protect intellectual property, implement cost-sharing schemes, and accommodate user-provided facilities. Workshops have been held in Tokyo, Osaka and Kyushu and solicitations for proposals have recently been initiated. Industry response to date has been rather slow, especially in areas outside Tokyo and Osaka [Reference 7].

4.3 Barriers Confronting Commercial Utilization of the JEM

There are a number of identified potential barriers to commercializing the operation and utilization of the JEM. A near term prevailing issue is the current economic conditions in Japan. Beyond this condition, however, most barriers involve the common themes of risk and cost, just as they do for their ISS partners. The following are representative concerns:

Lack of experience/knowledge about space environment. Most non-space industry companies are simply not aware of what possibilities the space environment holds. Conferences and workshops help but more are needed.

Lack of intellectual property protection. If companies cannot retain rights to intellectual property, they have no incentive. Universities and national laboratories must be sensitive to this when collaborating with private entities.

Excessively complicated process and long lead time for conducting experiments. According to ESA's MAP Programme, a turnaround of less than 3 months and "easy access/cost efficiency" are needed.

On-orbit laboratory designation. Official Japanese government designation of the JEM as a scientific facility has prevented its use for other purposes. The government has not yet indicated a movement towards privatization. As a government owned facility, the JEM cannot be used to enable a private entity to reap profits solely for itself.

Lack of industry-academia collaboration. Traditionally, Japanese universities were prohibited from receiving private funds for research. Regulations are loosening but the stigma remains and a large faction of purist academics are still opposed. Sawaoka argues that universities and non-profit organizations need to take the lead on collaborative research projects in microgravity science [Reference 8].

Lack of inter-agency cooperation. Having three autonomous agencies involved with JEM, problems arise. Japanese policy separates the promotion of the industrial use of space (MITI), the conduct of science (STA), and the role of academic institutions (MOE).

Corporate aversion to risk. In Europe, industry has indicated to ESA that a success rate of at least 50% is needed for it to commit to funding ISS experiments [Reference 7]. Japanese firms favor incremental improvements (e.g., inkjet printers) over untested innovations.

Hardware-focus. Most manufacturing companies are focused on producing hardware. System integration, life cycle support, turn-key technologies, etc., are largely foreign business practices.

Lack of entrepreneurial infrastructure. The "best and the brightest" work for large corporations. Venture capital is scarce and banks are averse to making loans without collateral (usually land ownership).

4.4   Potential Market for JEM Service Sector

In view of Japan's current JEM utilization promotion efforts and the various obstacles to commercialization highlighted in the previous section, a market for a small yet potentially high value-added service sector, targeted at overcoming some of these obstacles, may exist. For example, as part of JSUP's ARPPIUS and beyond, there are companies that could provide the following types of services to facilitate JEM utilization by industry.

•   Development and support of hardware and procedures for ground-based research and on-orbit experiments
•   Provision of know-how relating to space utilization
•   Provision and setup of on-orbit experiment opportunities
•   Implementation of these experiments.

Companies like Japan Manned Space Systems Corp., Space Engineering Development Co., and Advanced Engineering Services Co. are well-positioned to provide these and other types of barrier-breaking services. Currently, each of these companies is increasing its activities in the operation and utilization support for both ground and on-orbit research, although all three companies still derive a majority of their revenues from NASDA.

4.5   Recommendations for JEM Commercialization

The Federation of Economic Organizations (KEIDANREN) has, for many years, advocated the industrialization of space as the fourth infrastructure for commerce after land, sea, and air. In their policy report, KEIDANREN promotes the following five policies to enable the Japanese players to remain competitive in the global playing field [Reference 9].

•   Make public test facilities available for private use for a small fee
•   Develop new techniques for cost reduction and manufacturing efficiency and/or relaxation of regulations
•   Promote the responsive use of spin-off technologies
•   Facilitate the transfer of technologies maintained by NASDA
•   Continue to allow other government agencies to use the space infrastructure for satellite communications and Earth observations.

There are many paths, which NASDA could follow, to implement these recommendations. As with the other partners, clear policies which give the private sector confidence of stability must come first.

5. Gold Mine or Cash Cow?

Industry investors clearly say that if being involved in "commercial space" requires profits elsewhere to offset space losses, they will choose low risk profits elsewhere and omit the space losses. They feel that they must see a profit path for the space involvement. The only way in which this happens is for the costs to industry to be defined such that profits are achievable within a reasonable time, if they are to succeed at their undertaking. There is no need to try and build any artificial model with ISS. History is full of examples where governments have paid sunk costs and underwritten operations costs because they perceived (rightly or wrongly) a long term value to the enterprise or the potential larger scale economic impact evolving from expanded jobs from spin-offs. There are precedents when new frontiers are involved. Historically, exploration has been a public initiative. It has been very high risk with some promises of great rewards (most of which never occurred). What has occurred has been long term developments of value, and then only when the private sector became involved and focused on its business interests. The long term payoff for the ISS is yet to be determined. But the short term provides opportunities for the service communities to turn a profit.

Acknowledgements
The authors wish to express appreciation to Mr. K. Shiraki of JSUP and Mr. K. Nozaki of NASDA for providing background information and excellent comments.

References
1. Henry, E.: Japan's Space Industry as an Emerging Competitive Arena, *The MIT Japan Program Science, Technology & Management Report, Vol. 2, No. 1*, pp. 2-6, 1995
2. National Space Development Agency of Japan: *NASDA's Space Utilization Activity Overview*, March 1999
3. Ministry of International Trade and Industry: *Uchuu Sangyou Kihon Mondai Kondankai Houkokusho*, http://www.miti.go.jp/past/b60701h1.html. April 28, 1999
4. Japan Space Utilization Promotion Center: *Space Utilization Frontier Joint Research*, December 1995
5. Shiraki, K. and Kobayashi, T.: *A Study on Space Station JEM Utilization for Applied Research by the Private Sector*, Paper IAF-98-T.4.04, presented at 49th International Astronautical Congress, Melbourne, Australia, September 28-October 2, 1998
6. Japan Space Utilization Promotion Center: *Applied Research Pilot Project for the Industrial Use of Space*, April 1999
7. Japan Space Utilization Promotion Center: *JSUP News*, March 1999
8. Sawaoka, A.: *A Realistic Scenario of Japan for Strong Connection between Microgravity Researches and Ground-Based Hi-Tech Industries*, presented at the International Symposium In Space '98, Tokyo, Japan, September 21-22, 1998
9. Federation of Economic Organizations (KEIDANREN): *Wagakunino Uchuu Kaihatsu/ Riyou oyobi Sangyouka no Suishin wo Nozomu*, http://www.keidanren.or.jp/ japanese/policy/pol180.htm. April 28, 1999

ISS: Overview of Where Station is at and a Concept for Commercial Utilization

J. Worley, The Boeing Company, M/C HS-42, 2100 Space Park Drive, Houston, TX 77058, USA

e-mail: jeffery.worley@sw.boeing.com

Abstract
Having outlined the early stages of the life of the International Space Station (ISS), current U.S. ideas for reducing the costs of using it — and hence for encouraging its commercial utilization — are introduced.

1. Current Situation

With the joining of the Russian FGB and the American Unity node (see Figure 1), it can truthfully be stated that ISS operations and utilization are already underway. On 27 May 1999, the Space Shuttle was launched (STS-96), carrying provisions for the ISS. Further equipment to be sent into space to mate up with the ISS is currently at the Kennedy Space Center. Later in 1999, it is anticipated that the Russian service module will be launched from Baikonur.

Figure 1. The Russian module, with solar arrays, joined to the U.S. Unity node in low Earth orbit

G. Haskell and M. Rycroft (eds.), International Space Station, 227-231.

Three further Space Shuttle launches are planned in quite rapid succession. STS-92, with the Z1 truss, control moment gyros and Ku-band and S-band communications equipment is undergoing ground testing now (see Figure 2). STS-97, with the U.S. photovoltaic array, thermal control system, Integrated Equipment Assembly, and S-band communications equipment, will follow. The U.S. laboratory carrying five laboratory System Racks is the cargo aboard STS-98.

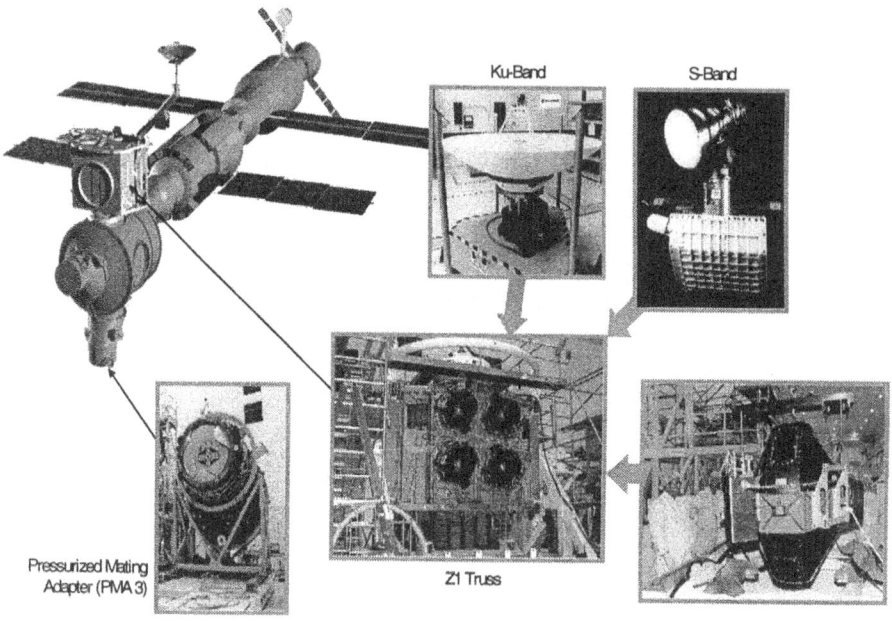

Figure 2. Equipment to be carried aboard the third American ISS launch, STS-92

Then STS-102 will carry aloft the Italian Multi-Purpose Logistics Module, named Leonardo, further Laboratory System Racks, the Human Research Facility for recording the health and performance of the astronauts, and the Integrated Cargo Carrier for delivering supplies and equipment to the ISS. The Leonardo module has already been tested at the Kennedy Space Center.

2. Concept for Commercial Utliization

The requirements of different commercial users of the ISS have to be balanced against the requirements for safely operating the ISS (which Boeing is interested in carrying out). Thus, the particular attributes of the present concept for using the ISS commercially include:

- an integrated environment which brings together the key needs of users and the requirements of ISS support
- complete support of the users' equipment aboard the ISS
- an effective interface between the customers and the ISS operator
- forward accountability for infrastructure systems and processes
- a clear pathway towards commercialization.

Effective — and streamlined — management is essential here.

Attempts are being made to reduce the costs of carrying out studies aboard the ISS, not only by seeking smaller costs for access to space but also by simplifying the complexity of ISS operations to make those more efficient. Key elements of the interdependent support and utilization systems are:

- program integration and support
- operations
- payload support
- selection of carriers and integration of cargo in preparation for delivery to the ISS
- product support.

Ways must be found to remove duplication and redundancies, multiple interfaces, and conflicting criteria. These should incorporate the industry's best practices and current technologies as much as possible.

The complementarity of the ISS operations' (O) systems and the users' (U) systems are illustrated in Figure 3. Here it is anticipated that a Non Government Organization (NGO) will be established [Reference 1]. The responsibilities of the NGO are shown as six bullet points within the circle on the left; DDT&E refers to Design, Develop, Test and Evaluate. The common, and recurring, elements of operations (O) and utilization (U) are shown in the central oval. The benefits expected to accrue from streamlining things are itemized in the box on the right of Figure 3.

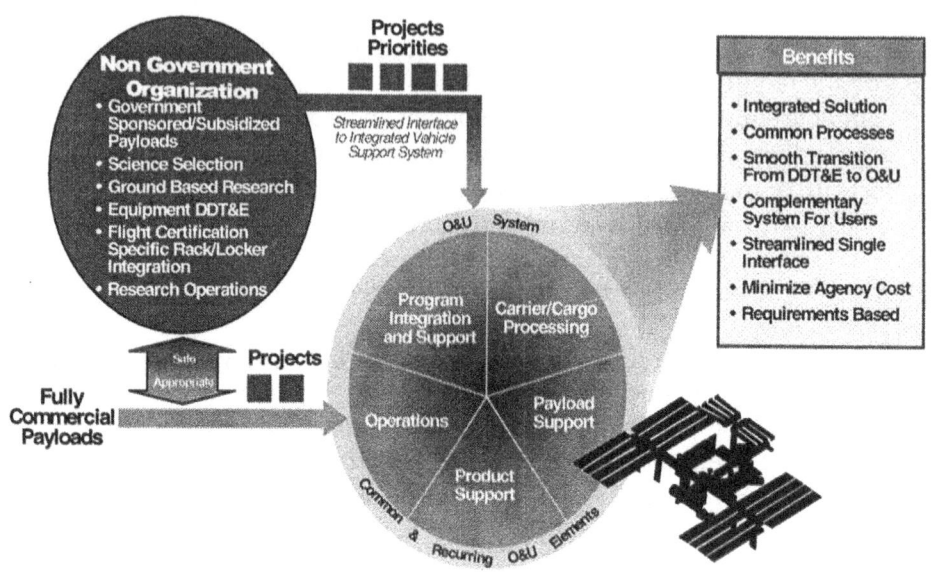

Figure 3. Common aspects of ISS Operations and Utilization, and the benefits to be gained by rationalizing these two systems

Further, this concept provides a natural pathway for commercialization and/or privatization of ISS activities, by allowing commercial utilization to evolve and by stimulating an international, market-based program.

3. Summary

Figure 4 illustrates the ways of reaching the goal that the operations and utilization of the ISS become economically viable, in order to "enable" the commercial utilization of the ISS. The left-hand side illustrates attributes which increase the cost of the ISS. These must be removed by conducting a dialog between those involved in the technology and the management of the ISS program, and by applying the "lessions learned" in earlier manned space missions such as Spacelab and Mir.

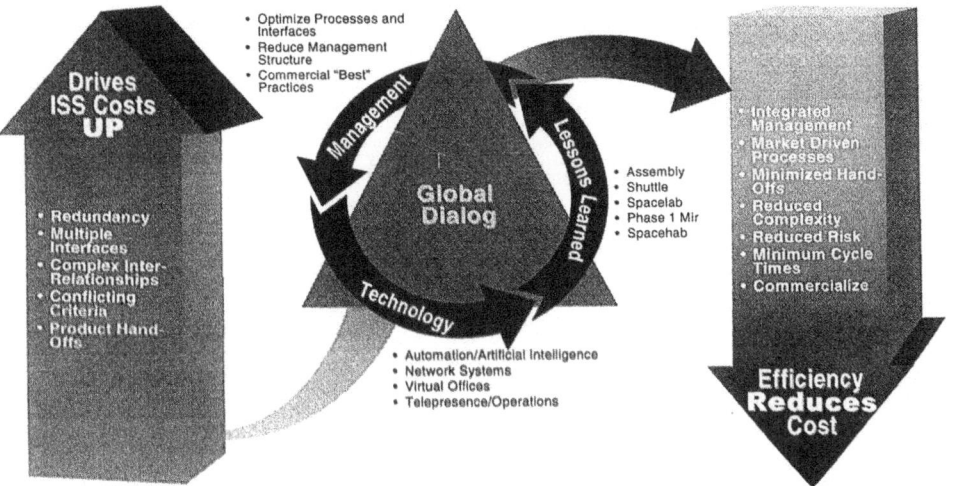

Figure 4. Diagram showing how the costs of ISS utilization can be reduced

Then the seven bullets in the right hand arrow will, by increasing efficiency, drive down the costs of using the ISS for the benefit of the peoples of the world.

References
1. National Aeronautics and Space Administration: Commercial Development Plan for the International Space Station, November 16, 1999

Report on Panel Discussion 5:

Technical and Management Innovations for ISS

R. Alexander, E. Benzi, International Space University, Strasbourg Central Campus, Parc d'Innovation, Boulevard Gonthier d'Andernach, 67400 Illkirch-Graffenstaden, France

e-mail: alexander@mss.isunet.edu, benzi@mss.isunet.edu

Panel Chair: Y. Fujimori, NASDA, Japan; International Space University temporary Faculty Member

Panel Members:

D. Beering, Infinite Global Infrastructures, LLC, USA
S. Gazey, DaimlerChrysler Aerospace AG, Germany
M. Uhran, NASA, USA
J. Worley, The Boeing Company, USA

Y. Fujimori led the discussion, which covered the need for a general business plan for the commercialization of the ISS, the role of small and big companies in the process, the problem of accessing the facility, and a long-term perspective for the the Space Station. He started by asking the panel members for comments on the usefulness of a business plan for the use of the ISS that would help the partner agencies to tune their commercialization efforts.

M. Uhran outlined the role of providers of information which the government agencies should maintain, and stated that it is the specific duty of the private investors to establish a business plan. The role of the government is not to select businesses, but to choose the best criteria to access the ISS facilities to ensure the success of the project, and thereafter step back. **D. Beering** pointed out that, in the communications sector, the private companies can do R&D for corporate goals, as well as establishing cooperation with NASA to provide services for the scientific sector. ISS is an excellent laboratory for commercial applications, as well as for research. **J. Worley** conveyed his corporate perspective, stating the need for a business case to justify the private sector's intervention, in a search of possible profitable markets. Servicing the ISS might itself be a good market — at least that is the Boeing belief — although "it is undefined as of today". Among the many hurdles to be overcome to ease the commercial exploitation of the Space Station, the biggest was identified as access to the ISS, both the long time required and the large launch costs. **S. Gazey** agreed on the existing difficulties, but put the emphasis on the necessity

G. Haskell and M. Rycroft (eds.), International Space Station, 233-235.
© 2000 *Kluwer Academic Publishers.*

for prompt and committed actions from the major private actors. The system itself should be utilized to learn how to develop the next generation technologies in order to provide more users with affordable and promising tools for their businesses. Optimization should be sought in all fields, from logistics to astronaut training, particularly, in the first stage, by the natural operators of the ISS, who are also the providers of the facility to the commercial world. These operators are, for their knowledge of the hardware and the consequent possibility to improve its design and operations, the agencies' prime contractors. The existence of a business plan that DaimlerChrysler Aerospace has been developing was outlined in this regard.

J. Cassanto (ITA), from the audience, directly addressed the Boeing representative by asking for the reasons behind the retreat of the company from its leadership position in the space privatization process. J. Worley underlined the difference between the past and the present with the merger having created the need for a reevaluation of the space business area. J. von der Lippe (INTOSPACE) stated his concern about smaller users with regard to the complexity of the interfaces. J. Worley agreed, and suggested a co-operation of the whole users' community to identify the necessary steps to improve the situation, one being the establishment of clear and easy to follow guidelines to access the ISS. S. Gazey underlined that the needs of the agencies and the large organizations necessary to operate the Space Station do not correspond to those of the small users. A specific effort is necessary to learn from the Mir example and offer to small users the shortest possible turnaround time and ease of access. The possibilities existing in the communications field were used by D. Beering to further stress the need for simplified interfaces. M. Uhran emphasized the universal desire to reduce the costs and time of access to the ISS.

An indication of the price which Boeing would charge for a new module (in comparison to its Russian counterpart, Energia) was invited by E. Dahlstrom (space development consultant); J. Worley pointed out the relative exiguity of the incremental cost, i.e. a possible big decrease in price. S. Gazey gave the example of the COF module and the halving of the original cost through reviews and building on experience. R. Moslener (Boeing), from the audience, clarified the will of Boeing to re-enter the private segment via co-operations with small companies or groups.

The problem of the time for access to the ISS and uncertainty of funding was posed again in relation to the short turnaround time for investigators in the microgravity science community. This question of C. Rousseau (ISU) was answered by M. Uhran with the consideration of the general availability of funding, while the resources to access the facilities are not enough. S. Gazey added the need to keep in mind other facilities beyond those in space, and noted the relative good state of the microgravity science sector in Europe. As a follow on to the question, the problem of a possible over-selling in the past of the microgravity market, and the possible re-establishment of a more realistic perception was raised from the audience. **M. Uhran** admitted NASA's and the other agencies' faults, and restated the problem of long access time as a major drawback. **Y. Fujimori** concurred, and also added that the science program is the most important for the ISS, and thus it should be made as attractive as possible.

J. Burke (NASA-JPL), with the help of a visual summary of the panel's main topics, asked for an identification of the practical steps to be taken to solve the problem of access to the ISS. **M. Uhran** pointed out the relevance of the users' guides as critical tools from which to start augmenting the possible gateways, which are becoming global, and are still undefined. Life science and microgravity should be included in the process, although their access methods are already good. **P. French** (ISU) questioned whether the need is for more gateways or for the present ones to be streamlined. **D. Beering** advocated the impossibility of following the commercial market, with its speed of change, as a motivation for the rules being changed within the market itself, and not, artificially, from outside. Greater simplicity was the solution proposed by **J. Worley**, as more gateways might just add complexity and cause chaos. He added that the present gateways allow for diversity; they should come closer together without becoming monopolistic. Of the same opinion was **S. Gazey**, who went even further by imagining a case-specific access for each commercial user, without too much importance being given to universal gateways.

The next question from the audience, posed by **C. Rousseau** (ISU), related to the long term vision (15 years ahead) of the panel members' involvement with the ISS. **J. Worley** forecast a scenario with a separate spin-off company operating the ISS independently of the rest of the Boeing company. **S. Gazey**, assuming an established commercial presence for the ISS, depicted DaimlerChrysler Aerospace selling on Earth technology and know-how acquired during the ISS lifetime and preparing for an international Mars mission. **M Uhran** pointed to the viability demonstration phase giving way to the real commercial exploitation phase, so that "the next Space Station will be private". **D. Beering** added his wish to see ISS being "considered as just another node of the world space network".

Y. Fujimori concluded by inviting all those involved to work together so that the public money spent on building the ISS could create a common business system.

Session 6

Concluding Panel

Session Chair:

K. Doetsch, President, International Space University

Report on the Concluding Panel

P. Messina, T. Brisibe, International Space University, Strasbourg Central Campus, Parc d'Innovation, Boulevard Gonthier d'Andernach, 67400 Illkirch-Graffenstaden, France

e-mail: messina@mss.isunet.edu, brisibe@mss.isunet.edu

Panel Chair: K. Doetsch, International Space University

Panel Members:

C. Bonifazi, ASI, Italy
D. Branscome, NASA, USA
A. Eddy, CSA, Canada
Y. Fujimori, NASDA, Japan
U. Merbold, ESA
H. Ripken, DLR, Germany

The concluding panel discussion was directed by **K. Doetsch** with the underlying theme that the symposium had been an analysis of the realities in utilizing the International Space Station, which could be described as a twentieth century pyramid, a stepping stone to space for mankind, or a market place. That analysis generated thought provoking comments such as "having developed the Space Station, the agencies should relinquish it". The debate commenced with a discussion concerning the long term prospects of the agencies acting as promoters of ISS utilization, and the validity of the need both to minimize government risk and responsibility and to increase the degree of risk and responsibility borne by the private sector.

   H. Ripken stressed that indeed the International Space Station is not a vehicle with a singular mission; the discussions have resulted in the emergence of the realization that it has many purposes, albeit with the need to confer responsibility as necessary. **Y. Fujimori,** who opined that NASDA holds a similar philosophy, supported this view. **C. Bonifazi** chose a diametrically opposed view by emphasizing that in fact the reality had already been finalized as the agencies would give preference to the scientific community; only with time and experience would a new community be formulated, thereby simplifying the interface between users and operators. Adopting a middle position **U. Merbold** stressed that it is the practice in numerous fields of endeavor to have government involvement and that the profit-driven motives of the private sector were crucial but had to be considered alongside other interests. **D. Branscome** pointed out that the private sector is necessary to

G. Haskell and M. Rycroft (eds.), International Space Station, 239-241.

improve activity in addition to reducing costs for ISS utilization. Considering the shortcomings of governments there is an obvious need to allow private sector participation in such a way that market-oriented decisions are the criteria on which to proceed, thereby leaving the market free to develop. **A. Eddy** buttressed this viewpoint further by stating that, except with political pressure from above, the agencies would not relinquish their ISS roles although economics could be a driver behind the politics.

K. Doestch then invited questions from the audience, the first of which was asked by **K. Rezkallah** implying that the developers of the International Space Station are selling something for which the customer is not yet prepared, thus creating a need to educate the commercial environment within the next five years. **C. Bonifazi** replied by pointing out that a lot of emphasis had been laid on construction of the ISS, but nonetheless that construction was an outstanding achievement in itself and would remain a greater challenge than ISS operation. **U. Merbold** disagreed, claiming that in the long run utilization of the ISS is more fascinating than its construction. **D. Branscome** chose a middle position believing that there is still a public awareness and educational need. This view was supported by **H. Ripken** who reckoned that scientific users are well aware of the possibilities offered by the International Space Station and a concentration on public awareness is necessary. A good demonstration is the overselling of the ISS as a platform for on-board experiments. The real challenge will be to attract commercial users and to sell the Space Station as a tool for solving the problems of industry.

A member of the audience was curious as to how liability and insurance issues would be addressed in the event of a force majeur. **A. Eddy** responded by pointing out the fact that all the partners of the Space Station cooperation are responsible and the cross-waiver provisions in the Intergovernmental Agreement addresses the issue. **Y. Fujimori** supported this opinion, confirming that the relevant rules are explicitly provided for in the said agreement and related events would be settled by discussion. **J. Cassanto**, inviting all the panelists to comment, reiterated the opening question posed by **K. Doestch**, claiming that the answer already exists. He believed that the NASA policy of devoting 70% of ISS resources to scientific activity and 30% to commercial activity should be maintained, and that the peer review policy needs to be abandoned with regard to commercial activity. **H. Ripken** stated that as far as Europe is concerned, by the authority of the European Utilization Board and the International Space Station User Panel, it had been agreed to divide utilization fairly between scientific applications, industrial and commercial needs.

M. Harrington then steered the discussion to a futuristic scenario by asking if the existence of entrepreneurs in 10 years that would put an end to users' subsidies would be considered as evidence of success of commercialization of the ISS. **A. Eddy** stated that 10 years is a long time as cost-based prices need to be lower than market-based prices in order for governments to recover their investments and the private sector to make a profit. **D. Branscome** felt that the concept of marginal costs would be used as a base, with policy and value costing being used as a means to recover costs. **U. Merbold** opined, with the concurrence of **C. Bonifazi,** that the results of utilizing the International Space Station would be non-material products — like knowledge — as opposed to physical goods, thereby making it difficult to come up with such numbers. **M. Rycroft** speaking from the audience simplified the question to: "what is the single best way, if not making a quick buck, making a slow buck with the International Space Station"? **D. Branscome** promptly replied..."if I knew, I would not tell you".

K. Doestch posed a last request to the panelists: what would you like to leave with the attendees and students? **C. Bonifazi** stressed that maintaining the proper attitude and speaking the truth was necessary. **U. Merbold** warned that the International Space Station is big and powerful, and that the scientific potential is not the most crucial issue because its justification was political — it is an outpost, and the realization of its scope will only emerge in retrospect. **D. Branscome** stated that the success of commercialization depends on its utilization and that the development of hardware in space is crucial. **Y. Fujimori,** labeling all in attendance as "space junkies", emphasized that marketing is crucial especially to the academic, research, institutional and industrial sectors. **A. Eddy,** addressing the MSS students in particular, advised them to refrain from underestimating what had already been done and to realize that they were pioneers who consequently needed to be daring, brave, innovative and creative in order to make the Space Station both a land of opportunity and a reality. **H. Ripken** suggested that co-operation between the commercial sectors in different nations would be difficult.

K. Doestch thanked the panelists for their views. He stated that there had been much talk in the presence of the custodians of the Space Station, and that he was much encouraged by the resulting interchange between the private sector, academia and governments. It was simply incredible that the International Space Station was becoming a reality for the next millennium.

Poster Papers

The International Space Station: Expanding our Knowledge of Reactions to Microgravity and Methods to Compensate the Effects of Gravity in Relation to Certain Species of Birds

L. Higgs, International Space University, Strasbourg Central Campus, Parc d'Innovation, Boulevard Gonthier d'Andernach, 67400 Illkirch-Graffenstaden, France

e-mail: higgs@mss.isunet.edu

Abstract
Since the launch of the first biological satellite, Sputnik 2, in 1957, experimentation has been performed to understand the effects on the anatomy of living and working in a microgravity environment. Many different species (examples being mammals and fish) have been used as test subjects, including humans; however, research using birds has been limited. This paper establishes the essential role that birds (Class Aves) should play in future space missions, how birds can be utilized to further other fields of research in relation to osteoporosis and other medical conditions, and why the ISS will be the most logical choice for this research.

1. Introduction

In the event of mankind inhabiting celestial bodies, the fragile ecological system associated with the Earth's biosphere should be recreated. Birds have an ecological niche in maintaining this equilibrium [Reference 1] and act as environmental indicators [Reference 2]. Their location in the food chain can show the health of their habitat as they feed on insects, thus controlling insect populations, and keep a balance between the organisms. Other key roles of birds include the reintroduction of nitrogen into the soil through their waste [Reference 2], seed dispersal and pollination of plants, detection of gases harmful to humans and, ultimately, a source of food (another level to the food chain).

Birds can also be subjects for numerous medical research studies. The skeletal structure of birds differs from that of other species as the bones are hollow and linked to a system of air sacs [Reference 3]. Research into the effects of prolonged exposure to microgravity may show an unknown effect. Other medical benefits may include research on conditions such as osteoporosis, infantile cortical hyperostosis (Caffey's Disease) and antitrypsin deficiency [Reference 3].

However, before birds can be introduced into the ecological system of a space-based habitat and ultimately benefit mankind, controlled experiments must take place to understand how they will react and adapt to their new

G. Haskell and M. Rycroft (eds.), International Space Station, 245-247.
© 2000 *Kluwer Academic Publishers.*

environment. The International Space Station provides the perfect environment for microgravity experiments and experiments with differing gravity levels, using the ISS centrifuge. Simulations can be performed at various gravity levels to mimic the conditions expected on a relevant celestial body.

2. The 'Life-Cycle' Project

2.1 Proposal

The proposed utilisation project would take place over the entire life span of birds to understand the effects of microgravity on their development. Beginning at hatching, the experiment would chart the development of the birds, through growth, into maturity and until death. It would be an ongoing project as it would continue with the offspring of the original test subjects. Constant research could therefore take place into how Class Aves adapt to life in space and how human development may also be affected.

2.2 Why Use the ISS Rather than Other Options?

The ISS is ideally suited to undertake this experiment due to the *controlled environment,* in which the progress of the caged test subjects is monitored. Multiple generations could be studied, to show any changes occurring over a longer period of time. Also the *long duration reduced gravity conditions* would help in readiness for life on the Moon or Mars. Experimentation on *confinement* could occur on Earth, but new effects could be expected aboard the ISS.

Other options include *parabolic flights;* however, these are only short duration flights. Although they would give valuable preliminary data, it would be difficult to show how birds react at different stages of their life cycle. *Shuttle-based experiments* would provide longer duration exposure to microgravity conditions, but insufficient time for the life cycle to take place.

3. Conclusion

The opportunities that will be made possible through the International Space Station are huge and will serve as an essential bridge between current understanding and greater knowledge. Stemming from a singular experiment, there would be significant spin-off possibilities available to both science and academia.

The ISS allows for biological research to be undertaken in an environment that will best resemble life on a celestial body. It will show how birds could live in space and adapt to reduced gravity conditions. Birds will be able to experience microgravity or reduced gravity conditions for a complete life cycle, and tests can be completed on them at different stages of their development. The Station could be used throughout its proposed life span (15 years), thus creating an experiment over many generations.

The possibilities of developing new cures for diseases as well as overcoming problems associated with microgravity are two key reasons for undertaking this research. From this, information on how life can develop in space from conception through to death can also be obtained.

A unique opportunity can also be envisioned from the applications available to education. Interactive experiments could be held between ground-based controlled experiments and the ISS, showing the differences between results. This could be wide ranging through its use by school, college and university students.

Creating a new, innovative and educationally stimulating role for the ISS could be achieved through the development of such a project. "...*Through individuals and groups theorising alternative and possible futures, they are analysing probable considerations, which will ultimately guide our development to other celestial bodies...*" [Reference 4]. This project will further our development and ultimately establish the International Space Station as an internationally beneficial tool to celestial colonisation.

Reference
1.   The Guinness Compact Encyclopedia, p. 140. Guinness Publishing, Enfield, 1994
2.   Earth   Generation: Facts   about   Birds, http://eelink.umich.edu/Curriculum/birdfacts.htm. January 28, 1999
3.   Encarta Encyclopedia: Birds, http://encarta.msn.com/EncartaHome.asp. January 28, 1999
4.   Higgs, L.: A Lunar Frontier: Comprehending Life on a Celestial Body, Master's thesis. International Space University, 1999

Diagnostic Solution Assistant (DSA): Intelligent System Monitoring, Management, Analysis, and Administration

C. Holland, Honeywell, Inc., Satellite Systems Operation: Space Systems, 19019 N. 59th Avenue, Glendale, AZ 85308-9650, USA

e-mail: cholland@space.honeywell.com

Abstract
The Diagnostic Solution Assistant (DSA) provides diagnostics of International Space Station hardware and software; the advanced Honeywell 'smart' model based technology performs real-time diagnostics. An astronaut or mission specialist can perform diagnostics of problems with Space Station electronic equipment and software on board the Space Station at any time. This feature provides 24-hour access to the knowledge of the best International Space Station hardware and software experts. The architecture of the DSA consists of a human-centered interface, diagnostic engine, system fault model, and system infrastructure.
The system is being developed in five phases.
Fault Isolation Phase. A user inputs fault symptoms to the system. The DSA uses the database to trace symptoms to possible sources, and presents the resulting list to the user, who then performs additional hardware or software tests to isolate the fault.
Fault Diagnosis Phase. The DSA performs fault isolation as in the first phase, and then displays an ordered list of inputs for the user to examine for faulty values. Also, the user receives instructions and test cases to assist in finding the faults.
Fault Detection Phase. The DSA monitors the signals of the system in order to detect symptoms automatically. The astronaut, mission specialist, technician, or engineer does not have to input symptom data. The DSA performs the fault isolation and fault diagnosis as in the first two phases.
Active Diagnosis Phase. The DSA monitors system signals and isolates faults as in the previous phases. Either the user may initiate the system by identifying a faulty output, or an error condition may trigger the diagnosis. The DSA executes automated test procedures to produce the smallest fault set possible before displaying the list.
Proactive Diagnosis Phase. This incorporates all the above, and applies advanced system monitoring techniques to predict or provide early detection of faults.

1. Problem Statement and Solution

The complexity of the International Space Station (ISS) requires that a full staff of ground-based system diagnosis experts is trained and available at all times. Response to critical situations must be immediate no matter what time of the day or night. Installation of new systems plus normal staff turnover cause personnel to be in training constantly. Domain knowledge lost due to staff attrition can never be regained. All of these factors lead to extremely high-cost ground-based flight system monitoring stations and sub-optimal efficiency. On-orbit diagnosis and recovery procedures are currently available only as massive binders of paper copies, making access difficult. Valuable time is wasted while the astronaut or mission specialist leafs through volumes of printed material.

G. Haskell and M. Rycroft (eds.), International Space Station, 249-250.
© 2000 *Kluwer Academic Publishers.*

The Diagnostic Solution Assistant (DSA) provides a solution to the inadequacies of the current on-orbit diagnosis and recovery plans. As fully integrated Vehicle Health Management (VHM), it offers a systems monitoring capability that is collocated with the system being monitored. DSA reduces the response time to complete the diagnosis, reduces the cost of maintaining a 24-hour manned monitoring center, and makes expert knowledge of systems operations constantly available. The DSA implements failure and fault diagnosis as an automatic function based on a system description database that captures the operational model of the system to be monitored. When a failure and fault occur, the response becomes an automatic function based on a system monitor procedure database, which captures the procedures to be followed. The database is updated as systems change and new systems are installed. Support is provided for on-line documentation access for flight crews, making paper copies unnecessary.

2. DSA Structure: The Five Levels

Fault Isolation Phase: Level 1 creates a tool that will capture system model information maintained by system experts to assist non-experts in fault isolation. The system model captures the operational description of the system. Fault information is input as states of system signals observed by the user who is maintaining the system being monitored.

Fault Diagnosis Phase: Level 2 enhances the user's ability to diagnose faults by providing direct access to system description drawings, specifications, and test procedures. Relevant documents are linked to specific areas of the system being diagnosed.

Fault Detection Phase: Level 3 monitors the system data directly by direct connection to the bus. The database for this level is modified to provide a description of the protocol to be used to collect live data and filter it for application to the system model.

Active Diagnosis Phase: Level 4 relieves the user of having to trace fault symptoms and conduct isolation and diagnostic testing. The DSA commands the system to collect more data to complete the diagnosis.

Proactive Diagnosis: Level 5 completes the VHM functionality by adding prognostics to the DSA system. System component failures are predicted to support scheduled preventative maintenance, and to facilitate planning for carrying replacement parts to ISS by the Space Shuttle.

Measurements of Raised Intra-Cranial Pressure, a Cause of Space Motion Sickness

C. P. Karunaharan, O. Atkov, International Space University, Strasbourg Central Campus, Parc d'Innovation, Boulevard Gonthier d'Andernach, 67400 Illkirch-Graffenstaden, France

e-mail: karunaharan@mss.isunet.edu, atkov@isu.isunet.edu

Abstract
Space motion sickness may be explained in terms of vestibular problems, fluid shift and raised intra-cranial pressure(ICP). The latter is known to trigger nausea, vomiting and headache in patients who have a defect in the drainage of cerebrospinal fluid. Different non-invasive as well as invasive techniques have been used in space to test this hypothesis. Solving the problem will enable astronauts aboard the International Space Station to be more effective.

1. Introduction

Different non-invasive as well as invasive techniques have been used in the space environment to test the hypothesis that increased intra-cranial pressure (ICP) causes space motion sickness. The former include Doppler ultrasound devices, tympanic membrane displacement devices and opthalmodynamometer. The invasive techniques so far include the Russian Bion experiments using primates which had probes placed intra-cranially prior to being flown.

The Variable Frequency Pulse Phase-Locked Loop (PPLL) measuring device developed by NASA [Reference 1] sends an ultrasound wave through the head, where it is reflected from the back of the skull and returned to a sensor. This device maintains a constant distance between the peaks of the outgoing and incoming sound waves by changing its wavelength. Therefore, with rising ICP, if the measurement between the front and the back of the cranium increases, so should the wavelength providing a marker for pressure. The first flight tests will be carried out during 1999. Alterations of ICP during acute 6° Head Down Tilt have been examined in humans using a non-invasive tympanic membrane displacement technique [Reference 2]. Early results point to a rise in ICP. In the volunteers studied, the stimulus intensity corresponded to a 20-25dB above reflex threshold, and at this level the mean displacement in the sitting position, +194nl, compared favourably with other investigations, 170 to 210nl. Tympanic membrane displacement of +194nl corresponded to an ICP of about 1.5mmHg. In 6 ° Head Down Tilt posture, a rise in ICP of 17mmHg was indicated. The opthalmodynamometry considered by Hamburg University [Reference 3] suggests a positive correlation between a rise in intra-ocular

G. Haskell and M. Rycroft (eds.), International Space Station, 251-253.

pressure in microgravity conditions (Spacelab D2-mission/German Mir Mission) with ICP. Using the automatic microprocessor controlled ocular tonometer the intra-ocular pressure at the moment of arterial collapse is measured. Clinical trials are yet to be done with this technique.

The invasive technique involved in measuring ICP by the epidural and subdural method was carried out on the primate 'Macata Mulatta' aboard BION ("bio-satellite") in 1995 [Reference 4]. Both in animals that are awake and under narcosis, ICP always rose when transferring into an anti-orthostatic position. However, it did not rise above the common "physiological" standard and there was no correlated arterial pressure change. If animals are kept in the anti-orthostatic position for many hours, there is a trend to ICP normalisation and, in fact, this is the case by about the ninth day of the flight The pattern of ICP pulse wave corresponds to the development of a considerable stagnation in the venous circulation of the brain during the fifth or sixth day of the space flight. Gradual ICP retrieval to the initial standard indicates adaptation of the system to the new dynamic conditions. For rabbits in simulated motion-sickness conditions [Reference 4] there was a pressure rise up to 15mmHg, whereas for those which were treated with lasex (1-3ml) a rise of only up to 7mmHg was noted. The second BION experiment was carried out in 1998 [Reference 5]. Within 5min of launch, the pressure rose to 13.78mmHg with an average of 10.23mmHg over the 20hr of the experiment. The ICP increases for the first 5 days, achieving values of 14mmHg. Unlike conditions on Earth, the rise during sleep is higher than when awake. Different results are obtained later in the flight. If ICP deviates from normal during sleep disturbances, reducing this may stabilise the upset sleep patterns which are a problem in space travel since they reduce work efficiency.

2. Proposed Experiment

ICP is potentially a critical parameter for understanding physiological changes (reduced blood flow and therefore decreased oxygen and metabolites) during actual and simulated microgravity conditions. A detailed understanding of the relationship between cerebral haemodynamics, cerebrospinal fluid and intracranial pressure changes in microgravity requires further studies. Clinical 'spin-offs' are of great value in neurosurgery, otolaryngology as well as in cardiovascular diseases. Our proposal is to use a combination of the methods discussed to measure ICP for a long period (perhaps on a weekly basis, on the ISS, with parameters such as EEG) to understand the correlation with sleep patterns and determine whether ICP increases enough (e.g. up to 7mmHg) to trigger sleep disturbances without any other clinical manifestations that are observed at higher pressures (up to 15mmHg). If this is the case, then controlling raised intracranial pressure will be an effective countermeasure to sleep disturbances and reduced working capabilities of astronauts aboard Mir and the ISS.

References

1. Nowak. R.: NASA's Space Biology Program shows signs of life, *Science,Vol. 268*, 1995
2. Hargens. A., et al.: Increased intracranial pressure in humans during simulated micro-gravity, *Physiologist, Vol. 35.* No1. Suppl.,1992
3. Draeger, J. A., Rumberger, E., Hechler, B., Linner, E.: *Non-invasive control of intracranial pressure rise due to fluidshift after entry into microgravity conditions using a new fully automatic microprocessor controlled Opthalmodynamometer*, Second European Symposium, "Utilisation of the ISS", ESTEC, Noordwijk, The Netherlands, November 16-18, 1998
4. Trambovetsky, E. V.: *Dynamics of the intracranial pressure of animals in zero-gravity model and in real space flight*, Airborne, Space and Marine medicine, Ph.D. Thesis, Moscow, Russia, 1995
5. Trambovetsky, E. V., Krotov, V. P.: Abstracts of Conference (Vol II), 10th Conference on Space Biology and Aerospace Medicine, Moscow, Russia, June 22-26, 1998

Recording Sprites, Blue Jets and ELVES from the ISS

A. Larisma, M. Rycroft, International Space University, Strasbourg Central Campus, Parc d'Innovation, Boulevard Gonthier d'Andernach, 67400, Illkirch-Graffenstaden, France

e-mail: larisma@yahoo.com, michael.j.rycroft@ukgateway.net

Abstract
Sprites, blue jets and ELVES are lightning-triggered optical emissions which occur in the rarefied atmosphere above thunderstorms. Images of these phenomena first appeared in the scientific literature less than a decade ago and the mechanisms responsible for their generation are not yet fully understood. One of the factors hindering a fuller understanding is the scarcity of good quality video observations around the world. Several features of the ISS make it an ideal platform from which to study such events. This paper proposes an experiment consisting of an astronaut tended, ground controlled, low light level video camera, sensitive to both visible and infrared light, mounted on an Earth-facing payload site aboard the ISS. Ideal observation windows would be selected beforehand to make optimal use of the data down-link bandwidth and astronaut time. This would involve using existing ground- and space-based meteorological observation networks to identify mesoscale convective systems (i.e. energetic thunderstorms), above which sprite phenomena typically occur. Astronaut observations from the cupola would facilitate the identification of regions of high activity; these are not always clear on wide-field video images. Real-time decisions made by the astronauts on where to point the camera would enable the operators to zoom in on an area of interest, and record events with high spatial and temporal resolution. Sub-millisecond timing of video frames, required for correlation with complementary ground-based observations such as very low frequency radio recordings and lightning detection data, could be supplied by the ACES clock on ISS. Coordinated ground-based, airborne and coarse satellite observations could be used for triangulation, and hence accurate position and size determination. Finally, the program could exploit the ISS's long lifetime by studying the effects of the solar cycle on the occurrence of sprite phenomena.

1. Introduction

Recent papers [References 1, 2, 3] have described the characteristics of red sprites, blue jets, and ELVES. Sprites are vertically aligned, often filamentary, luminous structures, extending from 35 to 90 km altitude. Their diameters may be as large as 50 km. The sequence of visible events preceding a sprite begins with a discharge in a thundercloud. After a delay of about a half-second, there is a brightening of the cloud. This is associated with a positive cloud to ground (CG) lightning discharge. The appearance of the sprite follows the discharge, with a characteristic development time of 10 ms. Additional discharges continue in the cloud after the sprite for about 1 s.

Blue jets are luminous cones that propagate upwards from cloud tops at about 100 km/s to altitudes of around 40 km. Unlike sprites, they are not associated with positive CG discharges. They typically occur in storm cells with

G. Haskell and M. Rycroft (eds.), International Space Station, 255-257.
© 2000 *Kluwer Academic Publishers.*

intense negative CG lightning activity and are usually followed by a decrease in this activity in a region extending about 15 km around the blue jet.

'Emissions of Light and Very Low Frequency Perturbations From Electromagnetic Pulse Sources', or ELVES, are large horizontal flashes that illuminate the sky for a region extending about 100 km around a causative lightning flash.

2. Proposed Experiment

An astronaut tended, ground controlled, video camera, sensitive to both visible and infrared light, mounted on an Earth-facing payload site aboard the ISS would be suitable for recording sprite phenomena. Recordings could be made in different modes depending on the temporal, spatial and spectral coverage and resolution desired. Different lenses, sensors and recording devices would be needed. Surveys of wide areas over long time periods could be made using a wide-angle lens and ordinary videotape. These could be used for statistical studies.

In depth views of the morphology of such celestial fireworks could be gained using high-speed video and a lens with a narrow field of view. Much of this footage would be of no value to the experiment as the chances of capturing an event, even under good observing conditions, are very low. The footage could be reviewed on board the ISS perhaps automatically, based on some signature of the phenomena, and uninteresting data could be discarded. This would help conserve data downlink bandwidth or storage medium bulk.

Observations could also be divided into astronaut-tended sessions and purely remote-controlled recordings. Survey recordings could be made with minimal interference from operators, while high-resolution recordings would benefit from astronaut intervention. Ideal observation windows would be selected beforehand to make optimal use of the astronauts' time. This would involve using existing ground- and space-based meteorological observation networks to identify mesoscale convective systems (i.e. energetic thunderstorms), above which sprite phenomena typically occur.

Astronauts should be able to control the camera, and have access to the video signal from the cupola. This would allow them to direct the camera based on visual observations. It should also be possible to relay the video and control signals continuously to the ground to enable scientists to control the camera remotely and make real time observations.

Observations of VLF waves, gamma rays or ~100 keV electrons could complement the video observations. Instruments could be specially installed for this purpose or use could be made of currently proposed sensors such as those on NASDA's Space Environment Monitor. Coordinated ground-based, airborne and coarse satellite observations could be used for triangulation, and hence accurate position and size determination. Observation campaigns could be arranged around one or two solar maxima and minima to investigate the dependence of sprite occurrence on the solar cycle.

3. Conclusions

Several features of the ISS make it an ideal platform from which to study sprites, blue jets and ELVES:

- The ISS's altitude provides a good vantage point from which to view large portions of the Earth
- The relatively low orbit of the ISS enables images with good spatial resolution to be made
- The inclination of the ISS's orbit allows for the major thunderstorm regions to be covered regularly
- The ISS's infrastructure can easily provide basic resources to the experiment such as power, communications bandwidth, accurate timing and cargo transport
- Existing instruments aboard the ISS such as NASDA's Space Environment Monitor can provide complementary data to the experiment
- Astronauts aboard the ISS can make complementary visual observations from the cupola, take decisions on where to point the camera, change recording modes or install new equipment
- The ISS's long life allows for observations to be made over one or more solar cycles.

References
1. Armstrong , R. A., Shorter, J. A., Taylor, M. J., Suszcynsky, D. M., Lyons, W. A. and Jeong, L. S.: Photometric measurements in the SPRITES '95 & '96 campaigns of nitrogen second positive (399.8 nm) and first negative (427.8 nm) emissions, *Journal of Atmospheric and Solar-Terrestrial Physics, JASTP, 60*, pp. 787-799, 1998
2. Boeck, W. L., Vaughan, O. H., Blakeslee, R. J., Vonnegut, B. and Brook, M.: The role of the space shuttle videotapes in the discovery of sprites, jets and elves, *JASTP, 60*, pp. 669-677, 1998
3. Wescott, E. M., Sentman, D. D., Heavner, M. J., Hampton, D. L., Vaughan, O. H.: Blue jets: their relationship to lightning and very large hailfall, and their physical mechanisms for their production, *JASTP, 60*, pp. 713-724, 1998

Gene Therapy as a Possible Counter-measure for Long Duration Space Flight

J. Maule, International Space University, Strasbourg Central Campus, Parc d'Innovation, Boulevard Gonthier d'Andernach, 67400 Illkirch-Graffenstaden, France

e-mail: maule@mss.isunet.edu

Abstract
Long duration spaceflight results in certain medical problems for astronauts such as bone weakening and a decline in the immune system. Whilst many methods have been used to combat these effects, they have been only partly successful. This paper proposes that spaceflight *counter-measures* be considered on a new and more molecular basis than has been done previously. One branch of molecular biology that has undergone rapid growth over the last decade is gene therapy. Experimental trials of this technique on the International Space Station could enable the development of an effective counter-measure for long duration spaceflights of the future.

1. Introduction

Gene therapy has been used on Earth for the treatment of cystic fibrosis, Duchenne's muscular dystrophy, Acquired Immuno-Deficiency Syndrome (AIDS), asthma and Severe Combined Immuno-Deficiency (SCID) (see Reference 1 for review). It is now being used extensively in human clinical trials [References 2, 3].

The simplest form of gene therapy involves a single injection of DNA into skeletal muscle, resulting in uptake and long term expression of foreign genes by muscle cells [Reference 4]. The procedure of injection and the events that follow intra-muscular injection of DNA have now been characterised in detail [References 1, 5, 6].

Many applications to spaceflight are envisaged, including administration of the *osteoprotegrin* gene [Reference 7] to prevent bone weakening, and stimulation of the immune system [Reference 1] to prevent decline of the immune system.

2. Methods

Plasmid DNA (circular DNA containing the desired gene) can be grown in bacterial cultures, isolated and purified ready for injection in less than a day (see Reference 1 for details). The plasmid DNA is dissolved in sterile saline solution and is then injected intra-muscularly in a similar fashion to conventional vaccine jabs. The method is simple, specific and inexpensive. DNA can also be stored easily at cabin temperature without degradation.

G. Haskell and M. Rycroft (eds.), International Space Station, 259-260.
© 2000 *Kluwer Academic Publishers.*

3. Conclusion

The long duration spaceflights of the future, such as a manned mission to Mars, will probably require biomedical counter-measures of greater efficacy than are currently available today. This paper has proposed that research should now be performed into new and non-traditional counter-measures. The substantial growth of gene therapy over the last decade should encourage experimental trials of gene therapy on the International Space Station.

References

1. Maule, J.: *The Characterisation and Modulation of the Immune Response following Direct Injection of Plasmid DNA into Murine Skeletal Muscle*, Ph.D. Thesis, London, Imperial College of Science, Technology and Medicine, 1998
2. Bolhuis, R. L., Willemsen, R. A., Lamers, C. H., Stam, K., Gratama, J. W. and Weijtens, M. E.: Preparation for a phase I/II study using autologous gene modified T lymphocytes for treatment of metastatic renal cancer patients, *Adv Exp Med Biol, Vol. 451*, pp. 547-55, 1998
3. Mackiewicz, A., Kapcinska, M., Wiznerowicz, M., Malicki, J., Nawrocki, S., Nowak, J., Murawa, P., Sibilska, E., Kowalczyk, D., Lange, A., Hawley, R.C. and Rose-John, S.: Immunogene therapy of human melanoma. Phase I/II clinical trial. *Adv Exp Med Biol, Vol. 451*, pp. 557-60, 1998
4. Wolff, J. A.: Direct Gene Transfer into Mouse Muscle in vivo. *Science, Vol. 247*, pp. 1465-8, 1990
5. Wells, K. E., Maule, J., Kingston, R., Foster, K., McMahon, J., Damien, E., Poole, A. and Wells, D. J.: Immune responses, not promoter inactivation, are responsible for decreased long-term expression following plasmid gene transfer into skeletal muscle, *FEBS Lett, Vol. 407*, pp. 164-168, 1997
6. Wells, D. J., Maule, J., McMahon, J., Mitchell, R., Damien, E., Poole, A., Wells, K. E.: Evaluation of plasmid DNA for in vivo gene therapy: factors affecting the number of transfected fibers, *J. Pharm. Sci., Vol. 87*, pp. 763-768, 1998
7. Cancedda, R. and Falcetti, G.: Mice Drawer System, presented at *Second European Symposium on the Utilisation of the International Space Station*, ESTEC, Noordwijk, The Netherlands, 1998.

Some Ideas for a Global ISU Educational Program Centered on the International Space Station

O. Zhdanovich, M. Rycroft, International Space University, Parc d'Innovation, Boulevard Gonthier d'Andernach, 67400 Illkirch-Graffenstaden, France

e-mail: zhdanovich@isu.isunet.edu, rycroft@isu.isunet.edu

Abstract
An ISU global education program linked to the International Space Station (ISS) is proposed in this paper. The aim of this program is to improve access by the general public to the results of research carried out aboard the International Space Station and thus to receive more support for space programmes.

1. Concept

The International Space Station (ISS) can be a unique classroom for space education and space-based education for students as well as for the general public. Recently, the space agencies have organised a special inter-agency group responsible for the educational program of the International Space Station. This group is now developing its working structure, etc., but no real project has been proposed. Separately NASA and ESA have their own educational initiatives as, for example, the NASA KidSat and ESA's SUCCESS project. The KidSat payload would it make possible for teachers and students to observe various phenomena from space on the Earth with a video camera. ESA's SUCCESS contest asks students from Europe to propose experiments for the ISS. The best proposals will fly on board the ISS and the winners will also be offered internships or thesis opportunities with European industries/universities.

The International Space University teaches students from all over the world in the peaceful uses of outer space. The ISU global education program linked to the ISS proposed here is aimed at raising interest among the general public in space-based education, in space activities and in pre-competitive research. This program which is based on the three ISU principles — international, intercultural and interdisciplinary — would make it possible for teams of students of all ages, from different countries and different disciplines, to analyse data obtained aboard ISS, in order to develop scientific ideas and obtain results, and also to consider the important issues of space commerce and public outreach. Dr. C. Welch, ISU UK Affiliate, comments: "an ISS-related nanosatellite program could offer education in the areas of remote sensing, space environment, space science and satellite systems engineering for schools and universities in many countries. Given its international network, ISU is perhaps uniquely placed to initiate such an undertaking".

G. Haskell and M. Rycroft (eds.), International Space Station, 261-262.
© 2000 *Kluwer Academic Publishers.*

People of different age groups and backgrounds would be involved: schoolchildren (especially underprivileged groups, e.g. disabled children, orphans), university students (Bachelor, Master and Doctoral students), school teachers, university staff, researchers, the general public, policy and decision-makers, and the private sector. The fields of study would be very broad, ranging from life sciences to microgravity, science and technology, and Earth observations. For school children and students, the program can be developed based on the principle of simplification from the doctoral to the school levels. Utilising the facilities and experiments aboard the ISS, the participation of university students and staff could be in various disciplines — biomedicine, production of new materials, science/engineering, and environment. The projects could start with a close co-operation with the Principal Investigators, and with astronauts and cosmonauts during their training, followed by their carrying out projects on board the ISS, and post flight processing. Competitions could be held amongst the participants within each class of people involved. The participants would be linked to ISU, to the network of ISU affiliated campuses and to national coordinators for this ISU/ ISS education program, as well as the public outreach and educational offices of space agencies. The participation of other nations, such as China and India, would be desirable.

2. Conclusions

Educational programs are directly linked to the public outreach aspects of space programs. Altogether they aim to improve participation by the general public in space activities and to receive more support for these. A few suggestions as to how this can be done, some based on the experience of educational and public outreach programs for the Mir Space Station developed in Russia this decade, are:

- Special competitions for university students and staff, analyzing the results of experiments performed aboard the ISS
- Development of space lessons from space; special education programs on TV, with special competitions for children of different age groups; special competition for journalists from magazines and TV/radio for the best coverage of space and best "space" journalist of the year [Reference 1].

References
1.   Zhdanovich, O.: *Visionary Strategic Planning for the Space Exploration and Resources Exploitation in the 21st Century*, ESA WPP-151, January 1999

Epilogue

Epilogue to the ISU Symposium on "ISS: The Next Space Marketplace"

H. Ripken, German Aerospace Center, DLR, Koenigswinterer Str. 522, 53227 Bonn-Oberkassel, Germany

e-mail: hartmut.ripken@dlr.de

G. Haskell, M. Rycroft, International Space University, Strasbourg Central Campus, Parc d'Innovation, Boulevard Gonthier d'Andernach, 67400 Illkirch-Graffenstaden, France

e-mail: haskell@mss.isunet.edu, rycroft@mss.isunet.edu

Abstract
This epilogue aims to report the main sentiments with which the attendees left the Symposium, although it cannot formally be said to represent the Conclusions of the Symposium. Two specific Recommendations for Future Actions are made, however.

1. Closing Thoughts

1.1 Preamble

At the end of this most interesting three day Symposium, no Committee or Working Group was established to generate a set of Conclusions arising from the wide-ranging discussions that took place at the Symposium. However, the first author had already taken the initiative during the Symposium to draft a Position Statement. This was circulated to a number of attendees, and a fuller, second draft was soon created with two recommendations. After the Symposium, the second author circulated that draft to all attendees, seeking their views on it and asking for suggestions for improvements and changes. The third author considered all the inputs that were received together with his own impressions, and generated a third draft on which this Epilogue is based.

What follows here, therefore, is a complement to the Symposium, rather than a Summary of it. It is presented in the hope that it will assist in carrying forward the discussions around the world on the crucial topic of ISS utilization and commercialization.

It has recently been announced that, from 13 to 15 June 2000, a conference will take place in Berlin, Germany, with the name "ISS Forum 2000: New Opportunities in Space". One objective of that meeting will be to convince the scientific establishment, and hence also public opinion, of the great value of the International Space Station.

G. Haskell and M. Rycroft (eds.), International Space Station, 265-267.
© *2000 Kluwer Academic Publishers.*

The International Space University hopes that the papers presented, and the discussions reported, in this book will act both as valuable reference materials for the Berlin meeting and also as a stimulus to further discussions, plans and actions.

1.2 Specific Statements and Suggestions

1. Early in the new millennium, and driven mainly by political considerations, the ISS will become an integral part of the global infrastructure, enriching technology development, scientific research, education and commercial business on the ground for the benefit of all peoples. Broad education and public awareness programs for the ISS are required in all Partner nations to ensure that the ISS program is as cost-effective, efficient and productive as possible; such programs should play specific roles, each having elements of public service (outreach) and commercial opportunities (e.g. market development).

2. There are new business areas and commercial opportunities where the ISS can be a valuable, or even a necessary, tool. A broad range of commercial opportunities for both traditional space industries and non-space industries should be exploited, to encourage a wide spectrum of activities in the future.

3. To promote ISS commercialization, partners from academia, industry and government should work together to reduce the ISS programmatic and policy constraints in order to stimulate viable commercial activities on the ISS. This requires clearly understood, and stable, policies on access to the ISS, e.g., project selection criteria, pricing, confidentiality, and intellectual property rights.

4. Management and support structures should be as "user friendly" as possible in order to encourage utilization of the ISS by commercial entities. At the same time these structures should enable, not limit, competition between potential users in different nations.

5. In the long-term, space agencies are not well suited to operate Space Stations or to manage commercial activities in space. They should act as promoters of space commercialization. While the risks resulting from new developments need to be minimized for investors, the responsibilities for utilization – and the associated risks – should be transferred to the private sector, as each commercial area becomes self-sufficient. ISS utilization needs to be an open process, with as much private responsibility as possible, and as little public responsibility as necessary.

6. In order to seed and incubate commercial ventures, public-private sector partnerships having adequate resources are an important step towards successful commercial ISS utilization.

7. Experience from past and present private sector initiatives in space, and from the commercial use and operation of space facilities, should be injected into ISS utilization plans.

8. Both private and public sectors should identify, discuss and prepare for the elimination of policy, cost and technical barriers to commercialisation of the ISS.

1.3 Specific Recommendations for Future Actions

1. International information interchange and cooperation are required to promote ISS industrialization and commercial utilization. An "ISS Commercialization Working Group", with participation going beyond the 5 ISS partners, should be established with agency, university and private sector representatives, to identify the way forward and make recommendations on, e.g., the standards which ISS commercial users should meet.

2. An international Symposium should be organized in the year 2000 to review the progress made in the last 25 years and the prospects for the next 25 years in all relevant areas of space activities. Both the theoretical basis and the empirical evidence to date should be summarized by the leaders in these space fields, with a view towards answering the questions: "what studies should be performed aboard the ISS?", "what are the next steps in the manned exploitation of space?", and "how should they be pursued?"

Annex

Utilization of the ISS, A User's Overview

Utilization of the International Space Station: A User's Overview

E. Benzi, B. Boardman, T. Brisibe, R. Gao, L. Higgs, C. Maredza, J. Maule, P. Messina, R. Mittal, M. Rezazad, International Space University, Strasbourg Central Campus, Parc d'Innovation, Boulevard Gonthier d'Andernach, 67400 Illkirch-Graffenstaden, France

e-mail: benzi@mss.isunet.edu, boadman@mss.isunet.edu, brisibe@mss.isunet.edu, gao@mss.isunet.edu, higgs@mss.isunet.edu, maredza@mss.isunet.edu, maule@mss.isunet.edu, messina@mss.isunet.edu, mittal@mss.isunet.edu, rezazad@mss.isunet.edu

Executive Summary

"Utilization of the International Space Station: A User's Overview" is a primer designed to give the reader a brief, but comprehensive, multidisciplinary look at the ISS. Although its focus is basic, it is also broad enough for the experienced reader to find new insight and be led by the references to whatever depth of information is needed. Its purpose is to engage prospective users in beginning with the Station's capabilities and ending with the first steps to be taken when participating in a mission.

This document arises from the literature review for the Team Project of the Master of Space Studies Class of 1999 at the International Space University. It is the result of a group effort by all thirty-eight members of the class and began as a much broader work in the fall of 1998. An exhaustive multi-disciplinary review of Space Station literature proved too cumbersome and the narrower focus of a "User's Overview" was chosen. Although it was written with the early scientific user in mind, the utilization climate has been rapidly changing in the direction of commercialization. If the project had run for another month, this document would have reflected more of this change in the direction of industrial users.

"A User's Overview" begins with a brief look at the historical and political development of the Space Station. Manned space platforms have been considered since late in the last century, but not until the 1970's with the Salyut and Skylab programs was the vision achieved. The Russians' Mir has been in operation for thirteen years and seems to have more than nine lives. The current Space Station began with a NASA Task Force in 1982 and was endorsed by President Ronald Reagan in his 1984 State of the Union address. Both the design and the list of Partners has undergone various changes. Canada, Europe, Japan, Russia and the United States now work co-operatively based on

G. Haskell and M. Rycroft (eds.), International Space Station, 271-273.

an Intergovernmental Agreement (IGA) and four Memoranda of Understanding signed in January 1998. As of May 1999, the first two elements of the International Space Station, the Functional Cargo Block and the Unity element, are on orbit. The Spacehab double cargo module is due for launch in a matter of days and four other assembly launches are scheduled in 1999. Many variables will affect the timing of the assembly sequence, but the complete ISS is well on the road to reality.

"A User's Overview" continues with an examination of the laws applicable to its utilization. The legal framework in which ISS will operate is often overlooked and this oversight is addressed. Underlying principles governing utilization as expressed in Article 9.1 of the IGA are presented. The percentage utilization rights of each of the Partners are given. The co-ordination of Partners' activities through the Multilateral Co-ordination Board is discussed, as well as cross waivers of liability and the issues of criminal and torts liabilities. Perhaps the most compelling legal issue surrounding the ISS in the immediate future will be intellectual property rights. Close communication among the Partners will be necessary to ensure that the legal impact on commercialization is positive.

Next follows a brief but comprehensive technical description of the ISS. Its characteristics, elements and the Partners responsible are listed in detail. Following a discussion of the space transportation systems used, the Station's major systems are described. The section closes with an examination of the natural and induced environments in which the ISS will operate.

Section five of "A User's Overview" examines the actual utilization of the ISS. First is the technical and engineering utilization complete with listings of space and Earth-based applications, both practical and "unrealistic". Following a discussion of the ISS as an in-orbit test bed for new technologies and its use as a manufacturing, assembly and servicing station is a realistic appraisal of the technological constraints on ISS utilization. Section five continues with utilization for the life sciences. The full range of laboratory facilities is discussed in great detail on a Partner-by-Partner basis. The potential for advances in the life sciences aboard the ISS is great; particularly valuable will be the impact on space medicine as the Russian tradition for research on the effects of long duration missions is expanded. The third part of section five is devoted to utilization in the physical sciences. Space physics and microgravity are discussed before the facilities and research interests are described Partner-by-Partner. The issue of constraints on the quality of microgravity is addressed again. Section five closes with utilization as applied to observing sciences. A discussion of astronomy facilities and projects is followed by similar material on Earth observations. Again, a Partner-by-Partner examination is presented; the section closes with a balanced appraisal and critical analysis of the observing environment aboard ISS.

Section six of the "User's Overview" discusses utilization of the ISS for those readers who are actually ready to participate in a mission. It begins by examining the management and planning process, continues with a discussion of the Announcement of Opportunity scheme for scientific experiments and concludes with a description of each Partner's approach to the administration of utilization.

"A User's Overview" continues in section seven by describing new ideas for the utilization of the ISS and offering recommendations for the future. These recommendations are listed as they apply to life sciences, physical sciences and observing sciences.

"Utilization of the International Space Station: A User's Overview" concludes with additional ideas for the future. Three ideas for utilization not addressed in the ISS design are discussed: use the ISS 1) as a base for other free-flying platforms, 2) as a jumping-off point for resource recovery and missions of space exploration, and 3) as a construction platform. The "Overview" concludes with some thoughts on commercialization and the need for a broad multidisciplinary approach to long range planning for the whole life of the ISS

Utilization of the International Space Station: A User's Overview

E. Benzi, B. Boardman, T. Brisibe, R. Gao, L. Higgs, C. Maredza, J. Maule, P. Messina, R. Mittal, M. Rezazad, International Space University, Strasbourg Central Campus, Parc d'Innovation, Boulevard Gonthier d'Andernach, 67400 Illkirch-Graffenstaden, France

Abstract

After a brief introduction to the historical and political development of the International Space Station, this report documents in detail the legal environment in which utilization must unfold. Then, following a technical look at the Space Station as a whole, currently planned utilization is examined critically in four broad categories: Physical Sciences, Life Sciences, Observing Sciences, and Technology Applications. Reference is made to a Web-based User Facilities table which gives a module-by-module rundown of facility resources.

Next is basic information on how to put an experiment aboard; in the section called "How to Utilize ISS." The report closes with broad ideas for the coming utilization of the ISS and offers recommendations for the future.

Preface

This overview is the result of the teamwork of the 38 graduate students in the Master of Space Studies (MSS4, 1998-1999) program at ISU. It began as a 'literature review' for the topic "Utilization of the International Space Station". Undaunted by such a broad subject, the class plunged into the task only to discover just how broad and complex it is. At the time, the work was organized by discipline into six major chapters. Although the result was interdisciplinary, it was not at all integrated or particularly coherent. The class learned how challenging it is to communicate with one voice when it is stirred by so many souls from 24 different countries. This product has evolved from that initial effort.

One of the things that became obvious to MSS4 while working on this "Overview" is that it has been produced at a unique moment in time. At one point, this work was called a "User's Guide". After dwelling on the name for a couple of days, it occurred to us that we ought to contact some users and get feedback on what they wanted in such a document. So two of the class members ventured forth by phone and e-mail to do some marketing research. That is when we discovered that our project had become somewhat obsolete before it was even conceived. Virtually everyone interviewed suggested that the piece be directed toward commercial users, where the new market emphasis is. If there had been another month and no other topics to distract us, a commercial guide is what we would have produced. In addition, one of those contacted in the research process pointed out that ESA already published a User's Guide. We discreetly changed the title.

G. Haskell and M. Rycroft (eds.), International Space Station, 275-355.
© 2000 *Kluwer Academic Publishers.*

One doubt that has bothered this group from the very beginning is that there are really no new ideas concerning the International Space Station. A lot of the suggestions in the literature have been around for years. Still, we have documented some of these because so many feasible, practical ideas fail to be put in motion. It is good to keep sight of them.

MSS4 continues to learn a lot on the way to finding one cohesive voice with which to communicate. If this "Overview" has helped the reader to use or further understand the ISS, we have succeeded.

1. Introduction

The material in this "User's Overview" is intended to serve as the Literature Review for the group thesis of the International Space University's Master of Space Studies Program for 1999. Although most reviews of this nature are "exhaustive," the topic "Utilization of the International Space Station" is so broad that the humans involved became exhausted long before the available resources. This effort was pared down in size by maintaining a very narrow editorial focus on only the issues which directly affect utilization and the ability of users to access the ISS. An attempt has been made to reach a balance between stand-alone disciplinary sections and integrated multi-disciplinary ones.

After a brief review of the historical and political development of ISS, a thorough examination of the legal environment surrounding the Station is presented. Even though "commercialization" and "privatization" are buzzwords in the space business overall, the impact of these trends on interlocking legal relationships that affect ISS is not often discussed. After an overview of the legal framework of the Space Station, the specific law governing utilization is explored in detail.

The Technical Description of ISS that follows is, again, the result of deciding the difference between everything known about the Space Station and everything that is related to its utilization. It is not always easy to tell the difference. This section should provide new insight for all readers regardless of background.

The following sections detail current utilization plans that are, of course, mostly scientific experimentation. This detailed examination of experiments is organized in the usual national space agency groups: Technology and Applications, Life Sciences, Physical Sciences and Observing Sciences. Presenting this material in a consistent manner proved to be difficult.

Basic information on "How to Utilize the ISS" is then presented. Next are some ideas for future utilization and finally MSS4's recommendations for the future of ISS. In addition to Appendix 1, which is an exploded view of ISS at configuration complete, Appendix 2 is a precis of a stand-alone Facilities Table which is produced more completely on the World Wide Web. Vital ISS data are reorganized in a way that not only eliminate duplication in writing but is also unique as well as useful to the user. That web site is:

http://mss.isunet.edu/~mss4web/tps/publications/fac_table/index.html

2. Historical and Political Development

2.1 The U.S. Decision to Build a Space Station

Scientists and novelists in the soon-to-be space faring nations had already dreamt of a permanent manned space platform almost 100 years ago; witness the Russian Tsiolkovsky and the American Edward Everett Hale with his Brick Moon. The dream became reality in the 1970's with the USSR program Salyut. In 1973 NASA put Skylab in orbit. The first permanently manned space station, Mir, has been operated by the Russians since 1986 and has doubled its expected six-year lifetime.

The 1984 announcement- In 1961 President John F. Kennedy chose to direct the National Aeronautics and Space Administration (NASA) toward a moon landing in lieu of a manned space platform. Despite this stated program goal, NASA continued to look into the possibilities of a Space Station. Skylab was a very valuable experience but the development of the Shuttle coupled with massive budget cuts in the late 1960's forced NASA to abandon plans for a space platform temporarily. Once the initial work on the Shuttle was completed, planning resumed on the Space Station effort. The Space Station Task Force was created in 1982. NASA's plans were given a green light in 1984, when President Ronald Reagan, in the Annual State of the Union address, called for the United States to develop the Space Station. Primary objectives of Reagan's decision were to retain America's preeminence in space, to encourage U.S. competitiveness and contribute to national pride. For strategic and diplomatic reasons, the United States' allies, Canada, Europe and Japan, were invited to join the project [Reference 1].

ISS and its evolution- The Station underwent several redesigns. At the onset of the program, the baseline design had a dual keel, a rectangular truss with living quarters and laboratory modules near the center. After the Challenger accident in 1986, a single truss replaced the dual keel in order to compensate for a lower Shuttle flight rate. Further revisions took place in 1988 and 1992. A major redesign occurred in 1993 that aimed at reducing costs as well as increasing international involvement. The newly redesigned Space Station program did include Russian parts and merged elements of the planned Mir2 with those of the planned US Space Station.

2.2 The Other International Partners' Participation

Europe: from the 1985 package deal to the current contribution- The ambitious Long Term European Space Plan, endorsed in 1985 by the ESA Council, included the Columbus program as the intended European contribution to the International Space Station. Strongly supported by Germany, the Columbus Program eventually gained French support. In exchange, Germany agreed to contribute to the development of a new cryogenic engine for the launcher Ariane and to the Hermes spaceplane. The Columbus program was composed of three modules (a pressurized one, a Free Flyer and a polar platform) and represented a considerable part of the ESA budget at that time. By the beginning of the new decade the political and economic situation had changed considerably in Europe. The ESA Ministerial Council then decided to redirect the Columbus program and to intensify international cooperation in order to reduce costs. The final decision, made in 1995 was for the European participation to be reduced to only the Columbus Orbital Facility (COF).

Japanese participation- Japan had already been approached by NASA to contribute to the post-Apollo Program in the 1970's. The USA renewed the invitation in 1982 through the Space Station Task Force. The decision to contribute to the Space Station project was made in 1985. Studies on the Japanese Experiment Module (JEM) and its functions and capabilities were conducted in the period 1985-1987. The preliminary design phase began in 1990. In addition to the JEM, Japan is providing the H-2 launcher and the centrifuge for the US Lab in exchange for its share of launch costs.

Canadian participation- The foundation for Canada's participation in the Space Station project was laid by successful cooperation with the United States on the STS (Space Shuttle) program [Reference 2]. This cooperation resulted in provision of the Remote Manipulator System (RMS), a robotic system known as the Canadarm. Canada's involvement in the ISS followed President Reagan's invitation in 1984 to participate as a full partner in the Space Station program. Canada is providing the ISS with the Mobile Servicing System (MSS), a sophisticated robotics unit that will be used to assemble and maintain the Space Station as well as maneuver equipment and payloads around it. The program is managed by the Canadian Space Agency (CSA).

2.3 Current ISS Status

In January 1998 the Inter-Governmental Agreement (IGA) was signed along with four Memoranda of Understanding (MoU) between NASA and each International Partner's Agency involved in the ISS, namely CSA, ESA, Government of Japan (GOJ) and RSA. These international agreements along with various implementing arrangements and the general provisions of international and national space laws constitute the legal framework necessary for the operation and utilization of the ISS.

The first element, the Functional Cargo Block (FGB) was successfully launched on 20 November 1998, followed a few weeks later by the Unity element. Docking of the two modules has been performed successfully. Testing and Stage Integration Reviews (SIR's) have been completed on the Service Module, the Z1 truss, the P6 array and the US Lab. Each of these is scheduled for launch in 1999, a year in which five flights are scheduled.

3. The Law Applicable to Utilization

3.1 Basic Concepts

The IGA provides in Article 2.1 IGA that *"The Space Station shall be developed, operated, and utilized in accordance with international law, including the Outer Space Treaty, the Rescue Agreement, the Liability Convention, and the Registration Convention."* It should be noted that the IGA has not defined a uniform set of rules that would apply only in or on the Space Station, but rather, in dealing with a number of legal issues which are fundamental to utilization, "it has established a link between: (a) the different parts of the Station, these being the flight elements provided by each of the partners, and the personnel, and (b) the jurisdiction exercised by the Partners on their own territory. In other words, the rules constituting the legal regime are aimed at recognizing the jurisdiction of the Partner States' courts and consequently allowing for the application of substantive national law in such areas as criminal matters, civil matters, including liability issues, and administrative matters which cover among other things the protection of intellectual property rights (IPR) and the exchange of goods and data" [Reference 3]. The implications and complexities that could arise from the effect of this legal regime on the aforementioned areas will be discussed more explicitly later in this section. The foundation for this basic rule or link is stated in Article 5 of the IGA which provides that " *each partner shall retain jurisdiction and control over the elements it registers....and over personnel in or on the Space Station who are its nationals"*. Jurisdiction therefore is fundamental to utilization and a direct consequence of the assimilation of flight elements to the territory of Partner State(s).

3.2 Utilization Rights

Underlying principles- The basic principles governing the utilization of the Station are laid down in Article 9.1 of the IGA, which provides that *"Utilization rights are derived from Partner provision of user elements, infrastructure elements, or both. Any partner that provides a Space Station user element shall retain use of these elements, except as otherwise provided for in this paragraph. Partners which provide resources to operate and use the Space Station, which are derived from their Space Station infrastructure elements, shall receive in exchange a fixed share of the use of certain user elements".* The provision takes into account the fact that partners may elect to include non-partners or private entities in their activities and thereby permits their inclusion provided the purpose is consistent with the applicable agreements, MoUs and implementing arrangements. This is subject to prior notification to all the partner's co-operating agencies as well as timely consensus between the said co-operating agencies.

In generic terms, the right to utilize the Station is related to the provision of user elements, infrastructure elements or resources, and the percentages allocated to the partners is stated in Article 8 of the MoUs which are discussed in detail later.

Provision and allocation of resources- The percentages of the rights to utilize user accommodations such as pressurized laboratories to be retained by the partner providing these accommodations is expressed in Article 8.3 of the IGA as follows: *" NASA will retain the use of 97.7% of the user accommodations on its laboratory modules, 97.7% of the use of its accommodation sites for external payloads and will have the use of 46.7% of the user accommodations on the European pressurized laboratory and 46.7% of the user accommodations on the JEM;*

-RSA will retain the use of 100% of the user accommodations on its laboratory modules and the use of 100% on its accommodation sites for external payloads;

- ESA will retain the use of 51% of the user accommodations on its laboratory module;

- the GOJ will retain the use of 51% of the user accommodations on its laboratory module; and

- CSA will have the use of the equivalent of 2.3% of the Space Station user accommodations provided by NASA, ESA and the GOJ".

In addition to the above there is also a sharing arrangement between the partners based on the allocation of Space Station resources. In this regard "an agreement was reached between the original partners and Russia based on the premise that Russia on the one hand, and the other partners on the other, retain utilization of their own contributions to the Station, and seek to offset only those items that cross the interface. By way of illustration, it was decided that for the purposes of sharing utilization the Russian Partner would keep 100% of utilization of its own modules, thereby recognizing that the infrastructure element supplied to the Space Station by Russia for its own benefit and that of the other partners would enable it to accumulate up to 100% of the utilization right on its own module. This means that the percentage agreed upon, on the basis of 100% within the entire Station, between the founding partners could be retained for the purpose of sharing available resources" (sic) [Reference 4]. The MoUs go further to provide the precise percentages to be allocated to each co-operating agency in Article 8.3.d as follows: "*76.6% will be allocated to NASA; 12.8% of the utilization resources will be allocated to the GOJ; 8.3% will be allocated to ESA; and 2.3% of utilization resources will be allocated to CSA*".

Barter and sale of Space Station resources- Article 9 of the IGA provides that the partners shall have the right to barter or sell any portion of their respective allocations. The terms and conditions of any barter or sale shall be determined on a case by case basis by the parties to the transaction.

Co-ordination of activities relating to utilization- The co-ordination of utilization activities is governed by a complicated set of provisions stated in Article 8 of the MoUs. Simply put, the MoUs establish a Multilateral Co-ordination Board (MCB) which will meet periodically over the lifetime of the ISS program or promptly at the request of any partner with the task of ensuring co-ordination of the activities of the partners related to the operation and utilization of the Space Station. The MCB comprises representatives of all the co-operating agencies and will be chaired by NASA with its decisions made by consensus between the agencies and without any modification of the rights of the partners as provided for under the MoUs.

3.3 Crew

The provisions relating to the crewmembers who will be responsible for the operation and utilization of the Space Station while it is in orbit are governed by Article 11 of the IGA and Article 11 of the MoUs, respectively. The Article addresses these issues among others: the prerequisite that crews will be trained in order to acquire skills necessary to conduct Space Station operation and utilization, and then subjects them to a code of conduct to be developed by NASA with the full involvement of the other co-operating agencies and approved in accordance with procedures followed by the MCB.

3.4 Liability Provisions

Cross-waivers in the Intergovernmental Agreement- The Outer Space Treaty of 1967 states in its Article VII the general principle that the launching state is internationally liable for damage. This principle has been elaborated in the 1972 Liability convention. Liability will be based on fault in the case of damage arising elsewhere than on the surface of the Earth to other space objects or to persons on board the space object, and it will be absolute in case of damage arising on the surface of the Earth or to aircraft in flight (Articles II and III). With regard to damage to third parties, the same principles apply, and the launching states shall be jointly and severally liable (Article IV).

Under the IGA, the essential provision concerning liability is that each partner agrees to a cross waiver of liability pursuant to which each partner state waives all claims based on damage arising out of Protected Space Operations (PSOs) [Reference 5]. These include all activities related to the Space Station except the development, manufacturing, or use of products or processes on the Earth developed as a result of activities in outer space [Reference 6]. The purpose of the waiver is "to remove from the liability equation any damage suffered by one partner as a result of the activities of another partner. The partners become self-insurers for their own property damaged during 'protected Space Operations'" [Reference 7]. This cross waiver shall apply to any claims for damage, whatever the legal basis for such claims for damage, including delict and tort (including negligence of every degree and kind) and contract, against the following:

- Another Partner State
- Related entities of another Partner State
- The employees of any entities identified above.

- However, the cross waiver of liability shall not be applicable to:

- Claims between a Partner State and its own related entity or between its own related entities
- Claims for personal injury or death
- Claims for damage caused by wilful misconduct
- Claims concerning intellectual property (for intellectual property a specific regime has been concluded in Article 21 of the IGA and is discussed further in a later section).

Criminal liability- The IGA does not contain an article covering the uniform exercise of criminal jurisdiction. Rather in its Article 22 it provides for each partner to exercise criminal jurisdiction over its own flight elements and over its own nationals on board, wherever they may be. *Inter alia* it identifies jurisdiction that can be exercised by a Partner regarding misconduct causing damage to flight elements or injuring a crew member; constitutes the treaty as a basis to proceed with extradition; and confirms the non interference between itself and the code of conduct to be developed pursuant to Article 11.

Liability for torts- Unlike criminal jurisdiction the Agreements do not contain parallel articles on civil jurisdiction and liability except between the Partners themselves and their related entities by the extensive cross-waiver clauses discussed herein before. The Agreement provides that Partners remain liable in accordance with the liability convention and because that convention covers liability for damage caused by space objects it does not embrace claims that result from injuries on board the Space Station itself. "The absence of tort jurisdiction therefore constitutes a major omission and a failure to address the issue of civil jurisdiction over persons on board the Space Station at an early stage of its development will cause the problem to magnify when the Space Station becomes operational and Space Station crews and accompanying personnel have no blueprint for the civil order on board"[Reference 8].

Since the Partners have opted to apply earthly rules to space, three major concerns regarding the settlement of tort claims immediately arise due to the multitude of conflicting laws that govern our planet and because of the territorial orientation of national law. These concerns are: determining the forum; selecting the law; and enforcing the decision. "Another concern is the sovereign immunity of partner states. One of the respondents in a civil action arising from a tort on board will always be one of the partners. This stems from each partner's responsibility for national activities under the Outer Space Treaty" [Reference 9]. The Agreement does not address the issue of state immunity from tort claims, again due to the cross waiver clauses.

There are other related issues. For instance, a decision is needed as to the adoption of a limited liability regime. The liability Convention sets no limit on recoveries for damage caused by space objects. Compensation therefore is to be determined in accordance with the relevant international laws and the principles of equity and justice. Limitation of liability is not only necessary to serve the needs of commercial practicality; it is required to promote international commerce, encourage private investment and initiative, and provide a cap on liability. This will afford a realistic basis for calculating risk and the procurement of affordable insurance.

Intellectual property- In order to address the protection of intellectual property (IP) we shall recall the principles as expressed in various legal instruments and the IP clauses for the IGA as stated in Article 21. The basic principles underlying national intellectual property laws may be illustrated by one of the most venerable texts on the matter, the United States Constitution (1787). Article 1, Section 8, paragraph 8 states: " *Congress shall have the power…To promote the progress of science and useful arts by securing for limited times to authors and inventors the exclusive right to their writings and discoveries*". All modern national patent laws are based on the same basic principles, giving rise to codification of Intellectual Property Rights (IPR) in the respective countries. A general definition of IPR emerges from a mosaic of national laws. IPR is the legal right, which is obtained, exercised, interpreted and judged according to nationally enacted legislation and ensuing case law. It is the right to forbid third party exploitation, or to allow the exploitation by license on terms dictated by the registered IPR owner or his designated successor. The scope of the protection of IPR is defined by the filed instruments, for example, by the claims of a patent. The geographical scope of the protection is that of the territory of the state which has registered the IPR. And the IPR has a limited lifetime, for example, twenty years after the filing date for patents [Reference 10].

It should be stressed that because the IGA does not constitute a homogeneous set of rules to deal with protecting IPR but rather tries to establish the necessary links between: (a) the different parts of the Space Station, these being the flight elements provided by each of the partners, and the Personnel, and (b) the jurisdiction exercised by the Partners on their own territory, the IGA is aimed generally at recognizing the jurisdiction of the Partner States' courts and consequently allowing for the application of substantive national law [Reference 11]. The provisions governing applicable law (including, but not limited to, IPR law) in the IGA therefore follow the "flagship principle" [Reference 12] as applied to vessels on the high seas, or aircraft flying over international waters. This is synonymous with the extra territorial jurisdiction exercised by sovereign states over their registered aircraft or vessels. In this regard Article 21 states "*…an activity occurring in or on a Space Station flight element shall be deemed to have occurred only in the territory of the Partner State of that element registry, except that for ESA-registered elements any European Partner State may deem that the activity occurred within its territory*". Because the European Partner State (EPS) is in fact made up of 11 member nations specific rules in addition to the above would also apply as provided for in Article 21.4. In effect this implies firstly that "judicial procedures with regard to a patent infringement case shall not take place in more than one European Partner State's court and secondly that recognition shall be granted in all European Partner States to a license if the latter is enforceable under the laws of any EPS" [Reference 12].

It is worth mentioning that the above provisions are not conclusive on the legal mechanism for the protection of IPRs. The IGA in Article 8.4 of the MoUs forecasts the development of new legal texts before the exploitation phase of the ISS, *"procedures covering all personnel, including Space Station crew, who have access to data"* to be developed by the MCB. In addition, Article 11 of the MoUs proposes the development of a code of conduct which will, among other things, establish physical and information security guidelines. Finally, Article 19.9 of the IGA enjoins the Partners through their cooperating agencies to establish guidelines for security of information which is of primary concern in the exchange of data and goods.

The issue of intellectual property rights may prove to be one of the most important challenges to face Partners in the future. International competition is rigorous and researchers are not always forthcoming on the subjects of shared credit and discovery. Remedies and preventative approaches should be worked out in advance through the review process at the international advisory committee level.

3.5 Exchange of Data and Goods

The provisions guiding the exchange of data and goods are laid down in Article 19 of the IGA. Pursuant to these provisions there is an obligation on each partner to transfer technical data and goods considered necessary to fulfill the responsibilities of the transferring Agency under the relevant MoU's and Implementing Arrangements. The Partners are further obliged to apply their "best efforts" to handle requests for the transfer of data and goods expeditiously. Because there is a need to protect data which are sensitive in nature for export control purposes and proprietary rights for purposes of confidentiality, the transfer of technical data and goods may be subject to a Marking procedure (which is intended to trigger particular protection with the receiving agency). This Marking procedure has the following features:

- It indicates specific conditions regarding use of transferred data and goods
- It indicates that data or goods shall only be used for the purposes of the co-operation and shall not be re-transferred to a third party without prior written permission of the furnishing Partner State [Reference 12].

3.6 The Future

Chairman H. Sensenbrenner, addressing the United States House of Representatives Committee on Science on October 7, 1998, emphasized the fact that "historically, governments have always dominated space activity. However, numerous benefits flow from the change of government dominance to commercial dominance including the emergence of entirely new markets within the commercial sector, involvement in new technologies, etc.. The vast potential of the commercial space industry has been constrained, however, by government regulations and laws that have not kept pace with the latest technology and changes in the market place". Perhaps governmental awareness of this state of affairs may have led to the recent promulgation of the Commercial Space Act of 1998 by the USA which in its Section 101 provides for the undertaking of a study concerning *inter alia....*"*the opportunities for commercial providers to play a role in International Space Station activities, including operation, use, servicing and augmentation*".

4. Technical Description of the International Space Station

The International Space Station is composed of pressurized and unpressurized elements, distributed systems and the associated ground elements that are provided by all Partners. This section provides an overview of these elements and systems.

4.1 General Description

A summary of the ISS characteristics at the assembly complete configuration is presented here, with Appendix 1 showing an exploded view of the Station:

ISS Characteristics at Assembly Complete		
Span, m		108
Length, m		74
Mass, metric tonnes		420
Maximum power output, kW		110...120
Pressurized volume, m³		1200
Atmospheric pressure, kPa		101.3
Operational Orbit	Altitude, km	370-460
	Inclination, deg	51.6
Attitude stability, deg per axis		±2.5
Crew, person		Up to 7
Data rate uplink, Mbits/s		72
Data rate downlink, Mbits/s		144
Data rate uplink, TDRSS, Mbits/s		120
Data rate downlink, TDRSS, Mbits/s		120
Expected lifetime, years		15 +

Table 1. Final ISS Parameters

The Partners, including Member States of the European Space Agency through ESA, and the USA, Russia, Canada, and Japan through their space agencies, are responsible for providing different elements of the Space Station. Table 2 shows the elements provided by each partner.

Partner	Elements
Canada through CSA	- Mobile Servicing System - Special purpose dexterous manipulator - ISS ground elements
ESA Member States	- Columbus laboratory, including basic functional outfitting - Automated transfer vehicle (ATV) to supply and reboost the ISS - ISS ground elements
Japan through NASDA	- Japanese Experimental Module, including the basic functional outfitting, - Exposed Facility and Experimental Logistics Modules
Russia through RSA	- ISS infrastructure elements, including Service and other modules - Research modules, including basic functional outfitting and attached payload accommodation equipment - Flight elements to supply and reboost the ISS - ISS ground elements
The US through NASA	- ISS infrastructure elements, including a habitation module - Laboratory modules, including basic functional outfitting and attached payload accommodation equipment - Flight elements to supply the ISS - ISS ground elements

Table 2. Different Nations' Contributions to ISS

The major elements of the Space Station are:

- Modules and nodes, which house essential systems and provide pressurized habitable environments and laboratories, and unpressurized areas for experiments and external payloads
- The truss, which is mounted on the Unity node and provides a connection between elements and external payloads and systems. It also houses the solar arrays, radiators, batteries, external payloads, and umbilicals
- Unpressurized modules and elements, which provide some service such as power production, payload support, etc.

- The mobile servicing center and mobile transporter, which will be used to remove payloads from the Shuttle cargo bay and transport them to other locations. The remote manipulator arm can carry payloads up to 128 tonnes, while the Special Purpose Dexterous Manipulator System is capable of performing more delicate tasks.

A summary of the major ISS elements, their expected launch dates, their availability for utilization and the general ISS configuration is presented in Table 3. Launch dates are provided as of May 15, 1999 and are based on the Revision D sequence and NASA's October 1998 Planning Reference. TBD listings are to be determined as flight plans are still under review. A new assembly sequence schedule is expected in June 1999 based on delays to the Service Module. Readers can find current information at http://station.nasa.gov/station/assembly/index.html [Reference 13].

Principal Pressurized Elements			
Pressurized Element	Launch date	Dimensions, meters	Function
Zarya Control Module (FGB)	1998	13 x 4.5	Provides initial propulsion and power until the service module is activated; after that, serves as backup and propellant storage
Unity Node-1	1998	6 x 4.5	Nodes provide six docking ports, external attachment points for the truss, and pressurized access between modules
Service Module	Fall 1999	14 x 4.05	Provides early living quarters, life support, communication, power, data processing, flight control and propulsion systems; later it becomes the functional center for the Russian segment (ROS)
3-person permanent human presence capability	Jan. 2000		Soyuz crew return vehicle docked to ISS
US Laboratory module	Feb. 2000	8.2 x 4.4	Provides equipment for research and technology development with 13 ISPRs; also houses the laboratory support systems and controls the US Segment
Node-2	TBD	7 x 4.5	
Japanese Experiment Module (JEM)	TBD	11.2 x 4.4	Provides laboratory facilities for materials and life sciences research; contains external platform, airlock, robotic manipulator, and separate logistics module. JEM has 10 ISPRs
Cupola	TBD		Provides direct viewing for robotic operations and Shuttle payload bay viewing
Russian Research Module #1	TBD	Details not available	Provides facilities for the Russian experiments and research, analogous to US Laboratory module
Node-3	TBD	7 x 4.5	
Russian Research Module #2	TBD	Details not available	
6-person permanent human presence capability	TBD		
Columbus Laboratory	TBD	6.1 x 4.4	Provides facilities for the ESA experiments and research and has 10 ISPRs
Centrifuge Accommodation Module (CAM)	TBD		Contains 2.5 meter diameter centrifuge for g levels from 0.01 to 2 g, and houses 4 ISPRs
US Habitation Module	TBD	8.2 x 4.4	Provides six-person living facilities

Principal Unpressurized Elements			
Unpressurized Element	On-Orbit Date	Dimensions, meters	Function
Space Station Remote Manipulator System (SSRMS)	Apr. 2000	17	Supports ISS assembly and Orbiter cargo bay loading and unloading, capable of carrying payloads up to 128 tonnes
Mobile Transporter	TBD		Provides structural, power, data and video links between ISS and Mobile Base System (MBS)
Mobile remote service Base System (MBS)	TBD		Serves as a stable base for the SSRMS
Integrated Truss Assembly (ITA)	Oct. 1999	108	Utilization starts after MBS commissioning, houses 24 payload adapters
JEM – Exposed facility	TBD	5 x 5.2	With 10 payload adapters, used for experiments and observations related to microgravity and space environment
Columbus External Payload Facility (EPF)	TBD		Houses 4 payload adapters, used for experiments and research related to microgravity and space environment

Table 3. The Principal Pressurized and Unpressurized Components of ISS

For an up-to-date version of the assembly sequence table and other information about ISS, the reader can refer to NASA's ISS website [Reference 13].

4.2 Crew and Payload Transportation and Logistic Carriers

Payload transportation and logistic flights are required through the assembly and operation phases of the Space Station. The main launch vehicles include:

- US Space Shuttle
- Russian Proton and Soyuz launchers
- European Ariane 5
- Japanese H-IIA launcher.

The US Space Shuttle will be the most used vehicle for transporting elements, logistics, crew, and payloads to the Space Station. There are two options for the transfer of cargo to the Space Station via the Shuttle:

- The Italian built Multi Purpose Logistics Module (MPLM) for the transportation of pressurized cargo to/from the ISS with a transport capacity of 9 metric tonnes. The MPLM accommodates 5 powered racks for refrigerators, freezers and active cargo, and 11 racks for passive payloads
- The Unpressurized Logistics Carrier (ULC) for transporting unpressurized and external payloads to/from the Space Station.

The Russian Proton launcher is used for transporting the Russian pressurized and unpressurized elements to the Space Station. The Soyuz rocket is used for delivering the Soyuz crew vehicle and the Progress-M1 cargo spacecraft. The Progress cargo spacecraft is used for pressurized cargo supply, attitude control fuel supply and orbital reboost [Reference 14].

The Automated Transfer Vehicle will be used for transporting cargo and for orbital reboost of the space Station and will be launched on an Ariane 5 launch vehicle.

The H-II Transfer Vehicle (HTV) will also be used for transporting cargo items to the ISS and will be launched on a Japanese H-IIA launch vehicle.

The X-38 lifting body reentry vehicle is also planned for use on the Space Station for emergency crew evacuation and return to Earth. In the early years of utilization, Soyuz will be the interim emergency crew evacuation vehicle.

4.3 Major Station Systems

Guidance, Navigation and Control (GN&C) System- The GN&C system provides for orbit maintenance, attitude determination and control, orbital maneuvers, and reboost and rendezvous operations. This system also distributes data on the Space Station's exact orbital speed, attitude and altitude to payloads:

- Guidance (where the Space Station is heading) tells the Space Station which route to follow. Guidance is accomplished by the Russian Orbital Segment (ROS) propulsion systems and monitored by Mission Control Center-Moscow (MCC-M)
- Navigation (where the Space Station is located in space) includes the functions of state determination, attitude determination, and pointing and support [Reference 15].

State determination provides the Space Station state vector (position and velocity vectors at a specific time). Two Receiver/Processor GPS sensors allow the Space Station to determine its Space Station vector independently, without ground support.

Attitude determination is achieved by a GPS interferometry technique combined with the data from two Rate Gyro Assemblies and a propagator algorithm.

For Pointing and Support the state vector, attitude and attitude rate data are passed to other Space Station systems. This subsystem also (a) calculates the targeting angle of the solar arrays, (b) calculates solar line-of-sight and line-of-sight vectors, along with rise and set times, (c) supplies data for measurement of the Space Station's center of gravity location and moments of inertia, and also adjustments for the location of dynamic systems like the Mobile Servicing System, and (d) provides GPS time for all the systems [Reference 16].

Control (how to get to a specific location and attitude in space) of the Space Station consists of translational control and attitude (rotation) control.

Translational control is achieved by reboosting the Space Station using the main engine of a docked transport cargo vehicle, typically a Progress M1. The Space Station is reboosted every three months to offset orbital decay from aerodynamic drag and to maneuver it away from orbital debris if necessary. The Automated Transfer Vehicle (ATV) also contributes to reboosting.

Attitude control is provided by the Russian Orbital Segment (ROS) propulsion system and the US non-propulsive attitude control system, which consists of four Control Moment Gyros located on the truss.

Electrical Power System (EPS)- The EPS is responsible for the production, processing, storage, and distribution of electrical power to the Space Station as well as providing an emergency power supply in case of failure in any part of the Space Station.

The Russian Orbital Segment (ROS) and the US Orbital Segment (USOS) provide uninterrupted electrical power for their own segments, as well as power sharing for all other Partners to support the ISS operations and user requirements. These two systems together produce up to 110 kW of power.

The USOS EPS is based on a distributed system design where primary power (160 V dc) is generated by two sets of solar arrays, mounted on the truss. Nickel-hydrogen batteries are used to store power for use while the Space Station is in the Earth's shadow. Primary power is converted to secondary power (124 V dc) and then distributed to the users. The USOS EPS provides up to 80 kW of power, of which up to 50 kW is allocated to the users [Reference 17].

The ROS EPS is based on localized system architecture where the elements have self-contained EPS with their own solar arrays which are located on several pressurized modules as well as on the Solar Power Platform. Nickel-cadmium batteries are used for power storage. The ROS EPS provides up to 40 kW power. The generated power (32 V dc) is converted to a lower user voltage (28 V dc).

If a voltage level different from the one mentioned above is required by users, it is the responsibility of the user to perform the voltage conversion. The maximum available power is 120 kW once the ISS assembly has been completed.

Each power channel is configured to supply power for particular ISS loads; to provide redundancy, however, the assembly complete design provides for rerouting (cross-strapping) primary power between various channels as necessary. An important note is that only primary power can be cross-strapped. After conversion into secondary power, flow through the distribution network cannot be rerouted. If there is a failure within the Secondary Power System, there is no redundancy, and the entire downstream path from the failure is unpowered. Instead, redundancy is generally determined by user loads.

Because of the high demand on power resources, specific scheduling might be needed for some experiments.

Environmental Control and Life Support System (ECLSS)- The ECLSS maintains a comfortable and habitable environment throughout the pressurized modules, provides water recovery and storage, as well as fire detection and suppression. The system supplies suitable amounts of oxygen and nitrogen, controls the pressure, temperature and humidity, removes carbon dioxide and other atmospheric contaminants, and monitors the atmosphere for the presence of combustion products. The ECLSS also collects, processes, and stores water and waste used and produced by the crew. Fire suppression and crew safety equipment are also provided through this system.

The ECLSS maintains an atmospheric pressure of 101.3 kPa with an oxygen concentration of less than 28 percent during 6-person human presence capability.

The ECLSS subsystems are:

- Atmospheric Control and Supply (ACS) provides oxygen and nitrogen to maintain the pressurized modules at the correct pressure and composition for human habitation. Oxygen for the Space Station is primarily supplied by the electrolysis of water through an oxygen generator called Elektron, and also by the Solid Fuel Oxygen Generator, which produces oxygen by an exothermal chemical reaction. The USOS has four high-pressure gas tanks that distribute oxygen and nitrogen to various parts of the Space Station. The Shuttle recharges these tanks. The ACS also provides gas support to other elements such as the Thermal Control System, Crew Health Care Subsystem, and payloads
- Atmospheric Revitalization (AR) ensures that the atmosphere provided by the ACS remains safe and pleasant to breathe. This subsystem monitors the composition of the Space Station atmosphere, collects carbon dioxide from the cabin, and filters gas contaminants and odors from the cabin atmosphere. This system must be capable of keeping the carbon dioxide level at about 0.5 kPa and under 1 kPa. Carbon dioxide is generated by the crew at a rate of 1 kg/man-day
- Temperature and Humidity Control (THC) circulates air, removes humidity, and maintains the temperature of the Space Station atmosphere. Three levels of circulation, inter module, intra module, and rack circulation, minimize temperature variations, ensure homogeneous atmospheric composition, and provide smoke detection capabilities
- Fire Detection and Suppression (FDS) provides smoke detection sensors, fire extinguishers, masks, and a system of alarms. There are two area smoke detectors in each pressurized module and one smoke detector in each rack. The smoke detectors are located in the air paths
- Water Recovery and Management (WRM) collects, stores, purifies, and distributes the Space Station's water resources. Collected water from the condensers of the THC subsystem, water from EVA activity, wastewater, and the products of the Urine Processing Unit are sent to the ROS for purification and refinement into "potable" water. Crew requirements are approximately 3 kg/man-day for drinking and food preparation and 4.5 kg/man-day for washing and personal hygiene. Waste water includes approximately 1.5 kg/man-day of urine and 2 kg/man-day of respiration and perspiration in addition to waste wash water [Reference 18].

The ECLSS also includes the Flight Crew system, which provides the crew with a safe environment and the basic necessities for life. The major constituents of this system are:

- Restraints and mobility aids support Intra Vehicular Activity (IVA) and personnel mobility and equipment restraint

- Portable emergency provisions sustain the crew in the event of an emergency and ensure the survival of the crew if the pressurized element is lost
- Housekeeping and trash management facilitates routine cleaning and garbage disposal
- Crew Health Care System (CHeCS) enables an extended human presence in space by supporting the health, safety, well-being and optimal performance of the crew
- Lighting systems facilitate productivity
- Personal hygiene equipment supports personal hygiene and metabolic waste collection
- Wardroom and Galley and Food Systems provide nutritional support; approximately 0.64 kg/man-day of ashless, dry basis food with a calorific value of 10500 to 11700 kJ is required. Current agreements state that NASA provides half the food and RSA the other half regardless of crew make-up. Most of the Russian food is ambient stowed (freeze-dried, low moisture, or thermostabilized) and prepackaged in individual serving packages. In addition, fresh foods are used to provide variety and alleviate boredom. This includes in-season fruits and vegetables, and kielbasa, a tasty Polish sausage. The US food (before assembly complete) is based on the Space Shuttle food system. The food is packaged for use specifically with the Service Module (SM) wardroom table/galley
- Crew privacy provides a private area for sleeping, changing of clothes and off-duty activities.

Thermal Control System (TCS)- The purpose of the TCS is to maintain Space Station elements, equipment, and payloads within their required temperature ranges. This system consists of passive and active thermal control subsystems.

Passive thermal control subsystems consist of multi-layer insulation, surface coatings, and paints, which basically isolate the elements and prevent over-heating or over-cooling conditions.

The active thermal control subsystem is provided by pumping coolant fluid in closed loop circuits to collect, transport and reject heat. Heat from heat generating devices is collected through cold plates and heat exchangers, transported by pump package assemblies and rejected through external radiators. In the USOS Active TCS, water is used as the cooling fluid in pressurized elements and ammonia is used in the external areas. In the ROS Active TCS, a mixture of water and ethylene glycol is used for pressurized elements and freons are used for unpressurized and external elements. The total heat rejection capacity at assembly complete will be up to 75 kW and the temperature within pressurized modules will be in the range 18-24 degrees Centigrade. Electrical heaters are also used in locations where other means of thermal control cannot be applied, for example, to prevent external TCS fluid lines from freezing.   No active thermal control is provided for attached payloads mounted on the truss.

Command and Data Handling System- The C&DH System, as the "brain" of the ISS, monitors and controls all aspects of the Space Station's operation by collecting data from onboard systems and payloads, processing the data with different software, and distributing commands to the appropriate equipment. The C&DH also distributes payload and system data to the crew and to the ground controllers through the Tracking and Data Relay Satellite (TDRS) system. This system consists of data processors, control and monitoring processors, crew interface computers, data acquisition and distribution networks, interfaces to systems and payloads, and different software.

The overall Space Station computer system is comprised of various partner computer systems: the US Command and Data Handling (C&DH) System, the Russian Onboard Complex Control System, the Canadian Computer System, The Japanese Data Management System, and the European Data Management System. The US C&DH System provides the " Space Station level control" software which aids in configuring systems for certain ISS operations. These operations are divided into different modes: standard, microgravity, reboost, proximity operations, external operations, survival, and Assured Safe Crew Return (ASCR).

The C&DH System relies on network technology and is composed of three components:

- The local data buses, which provide a low data transfer rate
- The local area networks (LANs), which provide a medium rate data transfer capability
- High rate data (HRD) links, which provide payloads with high data rate capabilities.

The crews have an interface, via their laptops, with all the data systems including payload data and communication lines with the ground. The interfaces for the crew laptops are similar and the data displayed on them can be displayed on the ground.

Payload users may develop their own application software for integration into the payload computers. However, payloads are required to use standard services and interfaces for commands and for all communications with the LANs and buses.

The C&DH System time distribution system also provides a stable frequency and time reference which might be useful.

Communication and Tracking System (C&TS)- The C&TS is designed to support Space Station operations and scientific research by providing audio, voice, and data communications with the ground, other spacecraft, and inside the Space Station.

The C&TS is specifically for:

- Two-way audio and video communications among crew members onboard the Space Station in the American segment and two-way audio during Extra Vehicular Activity (EVA)
- Two-way audio, video and file transfer communications with Flight Control Teams located in the Mission Control Center-Houston (MCC-H) and payload scientists on the ground
- One-way communication of experiment data to the Payload Operations Integration Center (POIC)
- Control of the Space Station by flight controllers through the reception of commands sent from the MCC-H and from the Shuttle orbiter
- Transmission of the both system and payload telemetry from the ISS to MCC-H and the POIC.

The uplink transmission capability is via an S-band system from the ground for Space Station systems and payloads, and the downlink transmission capability is via the Ku-band system.

The subsystems of the C&TS are:

- The Internal Audio Subsystem (IAS), which distributes audio throughout the Space Station
- The Video Distribution Subsystem (VDS), which distributes video throughout the Space Station and to external interfaces including fiber-optic analog video lines and the Ku-band for downlink
- The S-band Subsystem, which transmits voice, commands, telemetry, and files
- The Ultra High Frequency (UHF) Subsystem, which is used for EVA and proximity operations
- The Ku-band Subsystem, which is used for payload downlinks, and two-way video and file transfer.

Data and commands are transmitted to/from the Space Station through the Tracking and Data Relay Satellite System (TDRSS) to/from White Sands in New Mexico, USA. The data are then distributed through a combination of satellite and ground links. During short periods known as the "Zone of Exclusion" (ZOE) when there is no TDRSS-to-ground coverage, and during particular Space Station attitudes when there is no line-of-sight between the ISS and the TDRSS, the Space Station and the TDRSS lose communications. For at least 60 minutes in each 90-minute orbit, users are able to transmit or receive data. Another very short communications disruption occurs when communications are being handed over from one TDRSS to the other; this disruption lasts about two minutes during each orbit.

Communications via the Japanese (DRTS) and European (Artemis) data relay systems are also under consideration.

The JEM and Columbus Laboratory have video and audio systems that are compatible with the US system. Video and audio signals are digitized into data packets and multiplexed with other data for Ku-band downlink transfer. All the video, audio and data signals have time synchronization for proper time stamping and voice/data correlation. The Russian elements use the SECAM video standard and have no connectivity with the other Partner video systems.

Because of the high demand on communications resources, specific scheduling might be needed for some experiments.

The operation of the Space Station requires a diversity of information services. These include command and control services, payload support services, and automated information security services.

Command and control services provide for the interactive monitoring and control of the systems, elements, and payloads, as well as for the acquisition, processing, transmission, storage, and exchange of data among partners and payload operators and users. This includes the exchange of data between the Mission Control Centers, Payload Operations Integration Center (POIC) and other ground centers responsible for integration, planning, execution, monitoring and controlling of payload operations. Two manned flight control centers, one in Houston, Texas, USA, and one in Korolyov, Russia, operate constantly. Although the main center is in Houston, Korolyov has full control responsibility in a back-up situation.

Payload support services allow the remote users interactively to access, monitor, and control the equipment and payloads from their home institutions in pursuit of their experiments. This service also provides the users with ancillary data required for the meaningful interpretation of experiment results. Examples include orbital position and velocity, attitude references and standard time references as well as physical characteristics such as an element's temperature, gas component partial pressure, humidity, or external environmental parameters.

Automated information security services control the access to the information network and ensure the integrity and quality of the data on an end-to-end basis. The Space Station does not provide data encryption services for payload user data; users may encrypt their own data if they wish, however.

4.4 Environmental Considerations

Payload users should be aware of the effects of the natural and induced environment on their payloads.

Natural environment- The natural environment is the physical ambience surrounding of the Space Station, which generally remains unperturbed by the presence of the Space Station but which affects its performance.

The Earth's atmosphere produces forces and torques that disturb the motion and attitude of the Space Station. The atmospherically induced disturbances and accelerations determine the microgravity environment of the ISS. The atmosphere also affects the flux of the trapped radiation encountered by the Space Station.

Plasma influences the extent of spacecraft charging, affects the propagation of electromagnetic waves such as radio signals used for telemetry, and causes surface erosion. Plasma also induces electric fields in the structure of the Space Station as it moves through the Earth's magnetic field.

Charged particles can penetrate deep into the structure and after that may cause ionized radiation. This radiation significantly affects materials, chemical processes and living organisms. Charged particles may also cause temporary or permanent upsets in electronic devices or affect the propagation of light through optical materials by changing their optical properties.

Electromagnetic radiation originates from the Earth and the plasma surrounding the Earth, from the Sun and from outer space as well as from the ionosphere. Intense electromagnetic radiation could greatly affect the Space Station systems and payloads.

Micrometeoroid or space debris collisions may seriously damage the Space Station and its payloads. Shielding and shadowing can protect its critical elements.

Induced environment- The induced environment exists as a result of the presence of the Space Station, in low Earth orbit.

Control operations such as boosting, docking and undocking operations, gravity gradient, atmospheric drag, attitude and orbital maneuvers, movement of crew and robotics, operation of moving systems and other factors greatly degrade the microgravity environment of the Space Station. The level of microgravity is not the same for all frequencies of vibration. For any given frequency of vibration, there is a normalized level of microgravity. At a frequency level of 0.1 Hz or less, a standardized, planned level of microgravity can be maintained for at least 50% of the pressurized locations for continuous periods of up to 30 days and at least 180 days per year. The greatest disturbances ($\sim 10^{-3}$ g) occur during Shuttle docking and Space Station reboost. Microgravity quiescent and non-quiescent periods are scheduled in advance. The predicted microgravity quasi-steady acceleration levels are shown in Fig. 1; here µG represents 10^{-6} g.

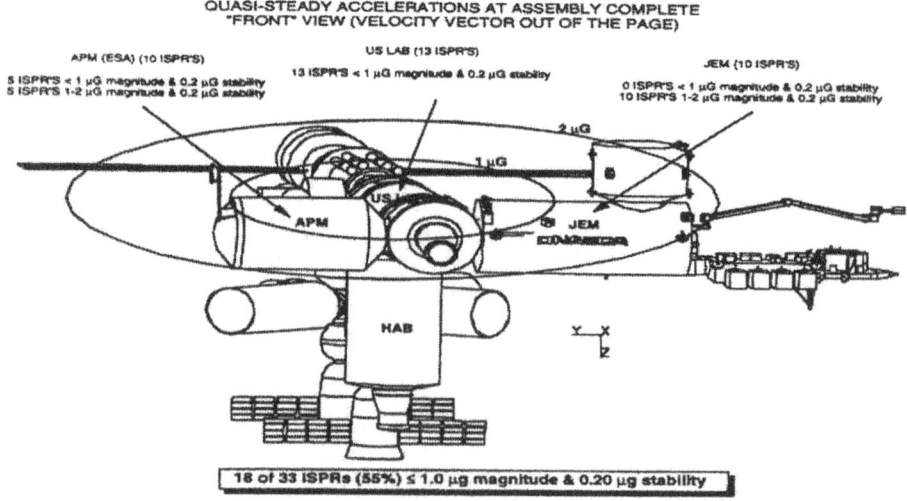

Figure 1. Microgravity conditions

The Space Station's external environment will further be affected by its presence, operation and motion, with induced effects due to:

- Plasma wake - the variation of plasma density from the ram to the wake
- Neutral wake - the variation of neutral gas density
- Plasma waves induced by the motion of the Space Station
- Vehicle glow on the forward, or ram, side
- Change of local plasma density and production of electrical noise caused by charging
- Enhancement of neutral density and the change of neutral composition by outgassing, offgassing, and the plumes from the thrusters
- Electromagnetic power radiated by systems on the Space Station
- Deliberate perturbation of the environment by active experiments and devices, such as (a) Transmitter/wave injectors, (b) Particle beam emitters, (c) Chemical releases, (d) Laser beams, and (e) Venting of any excess or waste fluids
- Visible light generated by the Space Station and reflections of sunlight from it
- Induced potential differences and currents that are generated by the motion of the Space Station through the Earth's magnetic field, which can draw current through the surrounding plasma.

5. Utilization

5.1 Introduction

This section focuses on what are likely to be four major areas of utilization onboard the ISS. These are technology applications, life sciences, physical sciences and observing sciences. In each of these four areas there is a description of the user facilities available, the potential activities accommodated by these facilities plus a brief critical analysis. A tabular description of a broad spectrum of user facilities on the ISS is given in Appendix 2. This appendix represents part of an ongoing facilities table maintained by MSS4 at http://mss.isunet/~mss4web/tps/publications/fac_table/index.html [Reference 19].

5.2 Technical and Engineering Utilization

As suggested by the complexity of the engineering system and the technologies involved (see section 4), the design, production of the elements, assembly and operations of the ISS have been and are, by themselves, an unprecedented and invaluable occasion for development of advanced engineering tools, methods and techniques.

Table 4 identifies, beyond the outlined technological test-bed nature of the project itself, the major areas of research in the engineering and technology fields, indicating some examples of possible applications of space-borne research to Earth [References 20, 21].

Space Application	Earth Application
Extra Vehicular Activity Support Systems	- Fire-fighting Suits - Toxic Waste Cleaning Suits - Deep Water Diving Apparatus - Cooling Systems for Physically Impaired Persons - Compact Power Tools
Environmental Control and Life Support System	- Waste Treatment - Environmental Clean-up - Agriculture
Structural Engineering	- Numerical Analysis of Structural Dynamic - Structural Verification Techniques
Aerospace Material/Space Environment Testing	- Lightweight O_2 Tanks - High Strength, Corrosion Resistant Pipes - Long Life, Self-Healing Paints - Lubrications - Solar Cells - Aerogel
Communications	- Support to R&D for Commercial Market - High Data Rate Power-Limited Networks
Information System (Radiation Exposure)	- Shielding & Hardening of Aerospace Electronic Hardware
Operations	- Warning & Emergency Systems in Confined Environment - Management of Pressurized Fluids - Controls and Interfaces
Fluid Management	- Multiphase Fluid Phenomena - Storable Fluids Fundamentals - Fluid Transfer - Free Surface Behavior - Thermal Non-equilibrium Processes - Cryogenics
Robotics	- Exploration of Unfriendly Environments (e.g. Ocean Depths, Volcanoes) - Hazardous Material Handling
Autonomous Systems	- Remote Installation Control - Safety and Reliability of Hazardous Facilities
Propulsion	- Low-Thrust Technology - Commercial Satellites Propulsion
Power Generation	- Solar Cells and Solar Dynamics - Batteries and Energy Storage Systems - Flywheels - Electric-powered Transportation Systems

Table 4. Space techniques and Earth applications

From an engineering point of view three broad areas of utilization are identifiable:

- Earth-oriented applications
- Test-bed for new technologies
- Manufacturing, assembling and servicing of the Space Station.

The proposed uses and experiments in these areas are very general, and are not easily linked to a specific facility on board ISS. The critical description of them, which follows, respects this characteristic.

Earth-based applications- Several projects are planned for the provision of Earth-based services:

- The global provision of precision time by means of a laser cooled atomic clock is foreseen with the Global Transmission Service (GTS). This service is pre-operational and the ISS will be used for test and demonstration purposes [Reference 22]
- A system for detecting and monitoring asteroids and comets that are potentially hazardous to the Earth is being considered for implementation on the ISS. Now that collision with comets 100 m or larger is recognized as a serious potential threat to the Earth's population, there is strong support for this activity from the EC. Similarly a system for monitoring space debris is being considered [Reference 23]
- The use of ISS as a research facility to study the effects of microgravity as well as a processing facility for the commercial production of unique materials needed by ground-based biotechnical, pharmaceutical, electronic and catalytic processing industries is identified in several proposals [References 24, 25, 26]. Commercial opportunities are expected for high temperature superconductors, CdTe crystals for X-ray detectors and zeolites which are extensively used in the oil refining industry to improve the quality of gasoline and diesel fuels
- As the microgravity environment inside the Space Station might not be sufficient for some of the very sensitive processes of interest, the servicing of free-flying platforms for the automated production of unique materials to be returned to Earth for further processing is proposed. Small scale free-flying platforms could also be used for inspecting the ISS structure with respect to damages caused by debris or for supporting EVA by providing artificial lighting [Reference 27].

Apart from these "realistic" applications, more visionary/speculative ideas for the utilization of the ISS (or any successor) are proposed. Some of these ideas will certainly have a significant impact on the overall configuration of the

Space Station and might not be compliant with the current plans for utilization (e.g. applications negatively affecting the microgravity environment or the Space Station configuration). Since in some areas the enabling technologies are still missing (e.g. cheaper access to space), the implementation of some of these ideas cannot be expected before the current design lifetime of the ISS is over, if ever. The majority of ideas are not related to a current, already defined mission.

One such "unrealistic" application, generating power by nuclear fusion processes, especially by use of a heavy isotope of helium (He3), is considered to be very attractive for solving the problems related to fossil-fuel or fission-based power generation processes. In this context the use of ISS (or successors) as an orbital docking station for spacecraft transporting He3 from the Moon is discussed. Compared to the Earth, He3 is abundant on the Moon and mining it is considered a viable option [References 28, 29]. The in-orbit trans-shipment of the He3 would be necessary as the atmospheric reentry can be more efficiently achieved with special reentry vehicles.

ISS: The in-orbit test-bed- The ISS can be used as an in-orbit test-bed for new technologies, as a promising alternative to the tedious on-ground qualifications typically achieved by a combination of more or less representative tests, simulations and analyses. Several areas of opportunity have been identified:

- According to some reports, the ISS has been proposed as a test-bed for several advanced methods for generating electrical power. These proposals aim at increasing the overall power generation efficiency or overcoming limits of current technologies
- ISS will be used to develop and qualify solar dynamic power generation systems for standard space applications. Compared to the conventional combination of photovoltaic solar cells and electro-chemical batteries, solar dynamic power generation, where the energy is thermally stored and converted into electrical energy via a thermodynamic cycle, provides higher efficiency (by a factor of 3). This leads to a smaller collecting area and less atmospheric drag in LEO than using solar arrays. This increase in efficiency is further enhanced by the lower losses of the energy storage system compared to current secondary batteries
- Another method discussed for generating power is the use of tethers. A long wire, which aligns itself radially to the Earth due to the gravity gradient, is moving in the Earth's magnetic field at the orbital speed of the Space Station, and thus inducing a large electric potential difference (voltage) between the ends of the wire
- The use of magnetically suspended flywheels for storing energy is another area under consideration. The advantages of such a system are a higher

energy density and practically unlimited lifetime compared to the restricted number of charge-discharge cycles possible with conventional batteries

- There are also suggestions to use the ISS for demonstrating the concept of solar power satellites where power generated in space is transmitted to Earth in the microwave band [Reference 30].

Other plans to use the ISS as an in-orbit test-bed are related to the following fields of technology [Reference 31]:

- Testing and verifying new propulsion systems (electric propulsion, in particular)
- Testing of new ECLSSs
- Testing of materials with respect to their radiation tolerance, their outgassing behavior and their degradation in the LEO environment (e.g. the Surface Effect Sample Monitor [SESAM] experiment)
- Testing of new electrical components, such as more efficient solar cells with respect to radiation tolerance and resistance to space debris
- Demonstrating precise deployment of structures or investigating their dynamic behavior (e.g. the High-resolution Photogrammetric Experiment [HIPE])
- Investigating the control/structure interaction in flexible structures and robots
- Testing new transportation systems for crew and equipment supply (ATV to ferry payloads to and from LEO, ATV-derived space robotic vehicle to assemble payloads in LEO) and Crew Rescue Vehicle being developed as a system between two merged projects [References 32, 33, 34]
- Reducing cost for the operation of space systems by testing autonomous on-board operation of spacecraft
- Demonstration of an optical communications link to augment services for payloads with a data rate up to 1 Gbps.

ISS: The manufacturing, assembly and servicing station- It is possible to intervene in such typical scenarios as failure to deploy a solar array or antenna, improper function of mechanisms, software errors, etc.. With regard to the assembly of spacecraft in space, the ISS is currently being considered for two missions.

- The X-ray Evolving Universe Spectroscopy (XEUS) mission requiring the in-orbit assembly of a mirror with 10m diameter [Reference 35]
- The next generation Very Long Baseline Interferometry space mission suggests the assembly of a 30 m radio telescope on, or close to, the ISS.

Assembling spacecraft in-orbit appears attractive as spacecraft design can be significantly improved if the component parts do not have to withstand launch loads. Today many parts of the spacecraft are designed just to survive the launch and ground transportation while in-orbit loads are negligible in comparison. Furthermore launching spacecraft in parts could allow better utilization of launchers. However, in-orbit assembly (in the current ISS configuration) requires dedicated EVAs which are a hazardous and inefficient way to perform the task. The small dimensions of the hatches on the ISS restrict its efficient utilization as an in-orbit satellite assembly and test facility, as these activities cannot be efficiently performed without EVAs. In this context it is also important to mention that the crew number is probably still too small considering the number of man-hours spent on the ground for the assembly, integration and verification of a satellite.

The ISS can also be used to maintain, re-fuel or repair satellites [Reference 25]. It is expected that a significant amount of money could be saved if failed satellites could be repaired in orbit, or the lifetime of otherwise perfectly operating satellites could be extended by refueling them or by exchanging degraded solar arrays, avionics, and payloads. This approach might, however, not be possible with current generations of satellites which are not designed for robotic or EVA servicing. Again, any intra-vehicular servicing activity would be limited to small satellites due to the small size of the ISS hatches and modules.

As for the use of ISS in support of future lunar or interplanetary missions, no detailed plans have been drawn up due to the lack of definition of such missions. Applications are conceivable in the long term for using the ISS in order to help develop enabling technologies for future manned missions to the Moon or to Mars. The idea is that the ISS could be used as an orbital staging point for equipment and supplies brought to the Space Station in multiple launches and assembled there for the outbound Lunar or Martian journey.

Technological constraints on ISS utilization- Two technological constraints limiting the broader utilization of the ISS can be identified:

- Low orbit, maintenance, and human activities have an adverse effect on the quality of the microgravity environment aboard the ISS. The quiescent periods, which should provide low levels of microgravity, have a maximum duration of 30 days. These periods are separated by either maintenance activities (10 days) or reboost maneuvers. However, the low microgravity environment may be disturbed, at any time, by atmospheric drag, mechanical vibrations, crew motion and possible debris avoidance maneuvers

- Considering the number of experiments to be performed on the ISS, the communication data rates are relatively low. The uplink data rate is only 72 Mbps, while the downlink capacity is 144 Mbps; when TDRSS is online, both uplink and downlink data rates will be 120 Mbps (see Table 1).

In the light of these limitations and constraints, the following upgrades can be proposed:

- The capabilities for communications with the ground can be augmented by use of commercial satellite communications services provided by LEO constellations
- The cost of access to space for many potential applications on the ISS is still too high, and only the development of new transportation systems can improve the situation. Quick access to space and an immediate delivery of the processed materials back to Earth has been considered crucial for the acceptance of the ISS by a broader scientific user community
- The development of an EVA suit with motorized hinges can increase the astronauts' efficiency and productivity during EVAs
- The attachment of a large diameter module with a large hatch would enable the efficient intra-vehicular assembly, repair, and refurbishment of satellites.

From the wide variety of possible technological uses of the ISS, both realistic and otherwise, we move to the very realistic business of ongoing scientific investigations. The International Space Station will soon be home to a rich variety of facilities dedicated to expanding Man's knowledge of himself and his environment. The rest of this section is divided into sections on Life Sciences, Physical Sciences and the Observing Sciences.

5.3   Life Sciences

Columbus- Much of this European module is devoted to research in the life sciences area.

European Physiology Modules (EPMs)- Current ESA plans are that the EPMs will contain the Advanced Respiratory Monitoring System (ARMS), the Advanced Bone Densitometer (ABDM) and the Bone Stiffness Measurement Device (BSMD), the Biomedical Analysis System (BMAS), the Station Off-Axis Rotator (SOAR), the Neuroscience Instrument, Cardiolab, and the ELITE-S2. Biolab, which occupies its own payload rack, is also discussed in this section [References 36, 37, 38].

The ARMS will be used for respiratory, pulmonary and cardiovascular physiology. The instrument is based on the photo-acoustic gas analysis technique that enables the measurement of oxygen, carbon monoxide, methane (blood non-soluble component) and sulphur hexafluoride gas concentrations. The ARMS will contain equipment for measuring blood pressure (Portapres), three-lead ECG (Ambulatory Electrocardiogram) and lung volume changes using a Respiratory Inductance Plethysmograph.

The ABDM measures the changes in structure and mineralization of the bone by using the propagation properties of ultrasound. The ABDM has evolved from the earlier EuroMir-95 BDM and therefore was designed to analyze the heel bone (calcaneum); the possibility of analyzing other bones on ISS is being investigated. The BSMD measures the longitudinal propagation of acoustic shock waves along the tibia (shinbone).

The BMAS will enable the onboard biochemical analysis of body fluids such as saliva, blood, and urine. Onboard analysis will reduce the need for frozen storage, provide quick turn-around of results and prevent possible degradation of stored fluids.

The Station Off-Axis Rotator (SOAR) will provide linear acceleration stimulus to the vestibular organs of its subject. On the STS mission Neurolab, SOAR was used in conjunction with an Eye Stimulation System (ESS or VEG) and an Eye Movement Recording System (EMRS or BIVOG); these may also be developed for the ISS.

CNES (France) and, formerly, DARA (Germany) have developed Cardiolab. It consists of the following instrumentation:

- Body Impedance Instrumentation — evaluates longitudinal body fluid shift changes
- Electrical Impedance Tomography Instrument — evaluates fluid volume changes across a body or limb section
- Near InfraRed Module Sensor Head — measures microcirculation
- EEG Module, with 8 electrodes- measures brain activity
- ECG Holter Module- measures cardiac activity
- Arterial Blood Pressure Holter Module- monitors pressure pulse at the wrist
- Portapres Blood Pressure Measurement Instrumentation- measures pressure pulse at the finger
- Blood Flow ultrasound Doppler module
- Air plethysmograph — measures absolute volume variations of limbs.

ELITE-S2 was developed by ASI (Italy) for tracking postural movement in three dimensions, using four infrared cameras.

Biology- Biolab will be used to investigate microorganisms, animal and plant cells, as well as small plants and invertebrates. The equipment of Biolab is contained in one ISPR and includes an incubator with centrifuge rotors. The Biolab Handling Mechanism provides an element of automation in sample processing. The Biolab Glovebox allows on-board sterile manipulation of samples.

Some of the equipment mentioned above and in subsequent paragraphs will be used on the ISS for studies in human physiology. These studies, together with human countermeasures, may produce vibrations detrimental to other experiments carried out simultaneously on the ISS that require very low levels of microgravity (e.g. fluid physics, observing sciences). The co-ordination between i) human activities and ii) experiments that require low microgravity needs careful attention. It is possible that a higher quality of scientific research on the ISS could be promoted by focusing on experiments not requiring a low level of microgravity rather than ensuring a selection of experiments from each of the scientific disciplines.

Protein Crystal Diagnostics Facility (PCDF)- The PCDF will be accommodated in the European Drawer Rack (EDR). It will be used for the same purpose as the APCF in the US HRF and the protein crystallization facility planned for the JEM. Although these facilities will perform similar functions (for example, producing highly regular crystals from protein solutions), the considerable demand for high quality protein crystals for X-ray or neutron analysis may justify the installation of three such facilities on the ISS. The elucidation of protein structures can lead to considerable advances in drug design and great benefits for patients on Earth. The additional factor of commercialization should also be taken into account. Each ISS Partner may have an interest in installing these facilities within his "own" module, as all experimental results obtained on ISS are under the jurisdiction (which includes patent law) of the member state who owns that module (see section 3).

The development of an instrument for in-orbit X-ray determination of the structure of processed crystals would produce faster results, save the expensive return of the sample to Earth and avoid the possible adverse effects of acceleration and/or thermal environments during reentry.

Cryosystem- This device is built by ESA under a NASA-ESA agreement and provides cold storage of biological specimens down to a temperature of -180°C. It consists of the Cryogenic Storage Freezer (CSF) and the Quick/Snap Freezer

(QSF). The CSF will store animal cells (from proteins to whole organisms), plant cells and protein crystals. The QSF will enable quick freezing of animal tissue and plant cells and snap freezing of animal organs, muscles and tendons, as well as of plant tissue.

The Minus Eighty degree Laboratory Freezer for the ISS (MELFI)-As the Cryosystem, this is built by ESA under a NASA/ESA agreement. It will store cell cultures, fluid samples and tissue samples up to 500ml in size. Freezing devices such as Cryosystem and MELFI generally create much heat. It is important that, during the planning of the TCS of Columbus, this has been taken into consideration.

The Space Exposure Biology Assembly (SEBA)- This will be used as an exposure facility for studies in exobiology and radiation biology [References 39, 40]. Exobiology is the study of extraterrestrial biology, i.e. the origin, evolution, and distribution of life. Some exobiology experiments use organisms that are adapted to growth and survival in extreme regions of the Earth's biosphere, such as microorganisms from desert or Arctic soil, airborne microbes from the upper layers of Earth's atmosphere, or archaebacteria [Reference 41]. Previous experiments have studied halophiles (a type of archaebacteria) in space. It was found that halophilic microbes could survive a two-week exposure to the harsh environment of space, within a certain type of porous rock. Other exobiology experiments are performed on molecules rather than organisms [References 42, 43].

EXPOSE- This is designed for experiments in photobiology and photo processing where samples are exposed to the UV radiation from the Sun. The experimental facility is a box-shaped structure that contains a tray mounted on a two-axis coarse pointing device. The tray contains multiple compartments, each of which has a lid that is controlled by a motor-operated system which enables precise control over exposure to solar radiation. Each compartment may be unsealed or sealed with quartz windows. Filters are also available, providing control over the wavelength of radiation incident upon the experimental samples.

Japanese Experimental Module (JEM)- The JEM contains various facilities which will be used for life science research: the Clean Bench (CB), the Aquatic Animal Experimental Facility (AAEF) and the Cell Biology Experimental Facility (CBEF). The CB will be used as a sterile environment in similar fashion to the LSG facility in the US CAM. Within the CB there will be a phase contrast fluorescent microscope and a monitoring camera.

Unlike the Aquatic Habitat of the US Gravitational Biology Facility, the AAEF facility will accommodate freshwater organisms such as Medaka fish, as well as saltwater organisms like Zebra fish. The AAEF, like the Aquatic Habitat, will be able to accommodate organisms for up to 90 days.

CBEF is essentially a cell and tissue culture incubator. It provides a controlled environment that is necessary for cell and tissue survival: 5% CO_2 atmosphere, 37°C and constant humidity. However, CO_2 levels can be varied from 0-10%, temperature from 15-40°C and relative humidity up to 80%. In addition, CBEF contains a rotating table that provides an acceleration up to 2g. The JEM also contains the Solution/Protein Crystal Growth Facility payload for protein crystallization.

Japanese Experimental Module Exposed Facility (JEM EF)- The JEM EF has 10 sites for Express Pallets. These offer further opportunities for experiments in exobiology and radiation biology.

Russian contributions- The Russian Space Agency (RKA) plans to contribute two scientific modules to the ISS, both of which may contain equipment for biomedical studies. The exact nature of the equipment on board these modules is not currently known. However, in the latter part of 1997 the Institute of Medical and Biological Problems (IMBP) in Moscow began to select proposals from Russian scientific and medical institutions for life science studies on the ISS. Fifty-eight proposals were selected — 23 in space physiology, 18 in space biology and 17 in the field of space flight medical support. Of these proposals, 41 are planned for execution during the early phase of ISS utilization (1999-2002) and 17 during the late phase (2003 onwards) [Reference 44].

In this case, the Russian approach to utilization has been to consult the scientific and medical research communities for proposals prior to defining the equipment that will be available. This strategy, which is more logical than the view taken by other agencies, where science proposals must be based on equipment already chosen by the agency and (in some cases) a relatively small group of scientists, may also promote a better quality of scientific research on the ISS.

As one of the five member Partners of the ISS, Russia co-operates with the other four partners in the area of life sciences via multilateral working groups, such as (a) the International Space Station Life Sciences Working Group (ISSLSWG), which focuses on many aspects of non-human life sciences research, and (b) the Human Research Multilateral Review Board (HRMRB), which co-ordinates scientific research on human subjects.

In addition, there have been reports of discussions between RKA and NASA regarding the salvaging of scientific equipment on the Spektr and Priroda modules of Space Station Mir for future use on the ISS. However, the RKA may decide to maintain Mir in operational use for a further 3 years (until 2002). As of early 1999, however, plans were underway to de-orbit Mir in a controlled fashion later in the year. NASA is pressuring RKA to devote its full energy to the ISS.

U.S. Laboratory- As with Columbus, much of this laboratory is devoted to the life sciences.

The Human Research Facility (HRF)- This is located in two ISPRs situated in the US laboratory [Reference 45]. Scheduled for launch in 2000, it will be used to investigate the cardiopulmonary, musculo-skeletal, neuro-vestibular and physiological systems of crew members. The HRF has a modular design facilitating changes of equipment within the rack. In addition to the equipment in the two ISPRs, there will be equipment located in the module aisle.

First Human Research Facility Rack

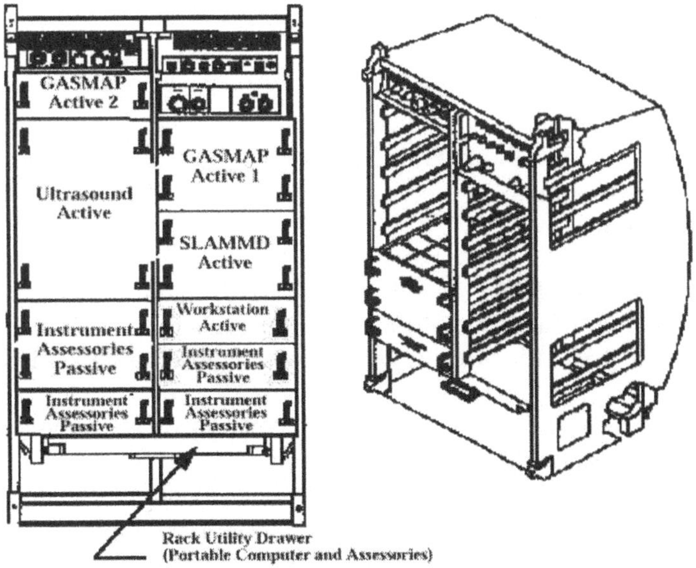

HRF dimensions (approximate): 6 feet (2 meters) high by 3 feet (1 meter) wide by 3 feet (1 meter) deep

Figure 2. The US Human Research Facility

In order to monitor many physiological symptoms, the HRF consists of a variety of equipment. First, the Activity Monitor, worn on the wrist, evaluates sleep patterns and daily activity. The Ambulatory Data Acquisition System (ADAS) uses physiological sensors to monitor core body temperature, blood pressure and respiration. In addition, there will be a Continuous Blood Pressure Device (CBPD) using a finger cuff for measuring blood pressure which can be worn for up to 24 hours. It should be noted, however, that there might be redundancy between ADAS and CBPD in measuring blood pressure. A third system is the Ambulatory Electrocardiogram (ECG), which will provide non-invasive, 24 hour ECG data. The Gas Analyzer for Metabolic Analysis of Physiology (GASMAP) and Space Linear Acceleration Mass Measurement Device (SLAMMD) are further US hardware. Equipment and a laptop computer will be available to collect samples and data and conduct on-board analysis. Non-US HRF hardware includes the Hand Grip Dynamometer (HGD) developed by ESA to measure hand strength and the Lower Body Negative Pressure device (LBNP) developed by Germany for cardiovascular studies.

Additional equipment for the HRF is under consideration, including:

- Bone Densitometer
- Core Body Temperature Monitor
- Fridge/Freezer
- Head and Body Tracking System
- Immunization Kits
- Skin Temperature and Blood Flow Monitor
- Eye Tracking Monitor
- Ultrasound device.

Centrifuge Accommodation Module (CAM)- This module will contain the Gravitational Biology Facility (GBF), Centrifuge Facility (CF) and the Life Sciences Glovebox (LSG) for carrying out research in gravitational biology.

Gravitational Biology Facility (GBF)- This consists of six habitats that can accommodate a variety of different living systems.

The Cell Culture Unit will maintain and monitor cell cultures for up to 30 days. The Aquatic Habitat will accommodate freshwater organisms, such as the much studied Zebra fish, for up to 90 days. Although the JEM contains the Aquatic Animal Experimental Facility (AAEF) and this may, at first glance, appear to duplicate resources, the two facilities have an important difference; AAEF accommodates saltwater organisms, but the Aquatic Habitat does not.

The Advanced Animal Habitat will be able to hold up to 6 rats or 12 mice. Compatible with this habitat is the Mouse Development Insert, which will house pregnant mice and their offspring during weaning. Many mammalian offspring died during weaning on Neurolab on STS in 1998. For there to be effective mammalian research on the ISS, greater research on improved maternal care and survival of mammalian offspring is needed during the weaning phase in microgravity. As a preliminary step, animal studies requiring astronauts could be carried out on US STS missions and those involving automated animal facilities as part of the Russian Bion program.

The Plant Research Unit supports plants up to a root-stem height of 38cm. Space studies on plant biology have examined every stage of plant development from seed to adult. Various plants have been studied in space including onion, Nigella, wheat, peas, corn, pine, beans, carrots, tomatoes, cucumbers, lettuce, mustard, spindle tree, barley, crepis, and Arabidopsis. These plants have been examined in terms of germinating capacity, survival, chromosomal disorders, radio-sensitivity, metabolic pathway disturbances, transport phenomena, chromosomal behavior, and reproductive development. The first plant to be cultivated through several generations in space (from seed to adult plant to seed, etc.) was Brassica rapa, a member of the mustard seed family. This plant successfully reproduced multiple times onboard Mir [Reference 46].

The Insect Habitat, a development of the Canadian Space Agency, will contain insects such as *Drosophila melanogaster* (fruitfly), which has been used extensively for studies in genetics and development [Reference 47]. Finally, the Egg Incubator will be used to incubate avian and reptilian eggs. All equipment of the GBF will be compatible with the Centrifuge Facility (CF) except the Insect Habitat and the Egg Incubator, each of which has its own internal centrifuge.

The GBF will be an important facility for the field of developmental biology. Developmental biology is studied in space in order to document the effects of gravity on the development of living organisms. There are many stages in the development of animals (including humans): oogenesis, oocyte maturation, fertilization, cleavage, axis formation, organogenesis (the point at which organs begin to develop within the animal) [Reference 48], and cellular and tissue changes throughout post-natal life. It is thought that gravity plays an important role in the normal progression of these developmental stages, although its exact role has not been completely defined. One experiment proposed for the ISS involves investigating a *Drosophila melanogaster* colony in space, over a prolonged period of time, in order to study the evolutionary changes and mutations of this species in space.

Centrifuge Facility (CF)- The CF will provide an acceleration of up to 2g for biological samples. In addition to the CF, the CAM will contain laboratory support equipment such as:

- 80°C Freezer
- 4°C Refrigerator
- Cryo Quick/Snap Freezer
- Passive Dosimeter System
- Dissecting Microscope
- Compound Microscope.

All this equipment is essential for biological research. Freezers, incubators and centrifuges, however, generate vibrations and heat. It is important that these items be tested for the vibrations and heat which they generate when in functional mode, as it has important implications for all microgravity experiments on the ISS.

Life Sciences Glovebox (LSG)- This is a sterile compartment where two crewmembers can manipulate biological specimens and conduct experiments at the same time by wearing gloves that extend into the 0.5m^3 volume of the LSG.

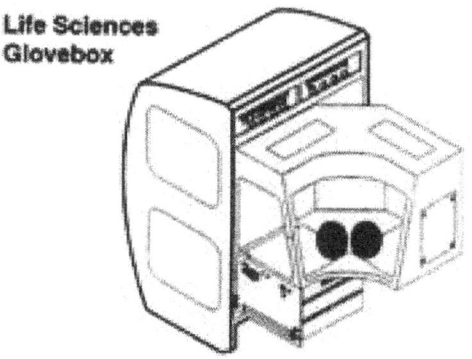

Figure 3. Diagram of the LSG apparatus

Two overviews of life science research in space have been published in recent years [References 49, 50].

5.4 Physical Sciences

Space physics- Space Station platforms provide two basic capabilities for space physics experiments [References 51, 52, 53]:

- Continuous monitoring of the "solar constant" (the total irradiance from the ultraviolet to the infrared), and the solar spectrum
- Continuous monitoring of the Space Station environment; some effects which the ISS has on its environment are summarized in section 4.4.

Several experiments have been proposed on the ISS to study the long term variability of solar properties:

- NASA has received a proposal for a High Resolution Solar Observatory
- ESA is considering proposals for a Solar Variability and Irradiation Monitor [Reference 54], long term measurements of Solar Diameter Variations and Solar Spectral Irradiance and Variability Measurements [Reference 55]
- Japan has proposed a Solar Radiation Monitor and Solar X-ray Imager [Reference 56].

A NASDA payload, the Space Environment Monitor, will examine the external environment near the Japanese Experimental Module (JEM). It contains a neutron monitor, dosimeter and dust collector. The Space Environment Monitor is, however, useful to quantify ISS environmental conditions for use by other experimenters and as a source of scientific data in its own right.

Microgravity Science- In microgravity, various phenomena are significantly altered, especially convection, buoyancy, hydrostatic pressure, and sedimentation.   Scientific disciplines affected include fluid physics and transport phenomena, combustion, crystal growth and solidification, biological processes and biotechnology. Microgravity may therefore be regarded as an important tool for increasing precision in the measurement of thermophysical properties, thereby improving models of complex phenomena and hence manufacturing processes on Earth.

ESA- In 1995, ESA member states prepared for the 'steady-state' utilization phase after assembly of the ISS and stipulated an ISS Utilization Preparation Program including a Microgravity Applications Promotion (MAP) program with three projects. With the latter program in place, ESA hopes to attract researchers from industry and academia and to develop relevant projects demonstrating the uniqueness of ISS as a tool for industrial research [Reference 50].

- Crystal Growth of Cadmium Telluride (CdTe)- CdTe materials are used in highly sensitive detectors and photorefractive devices. CdTe X-ray and gamma ray detectors have high potential in dental imaging and mammography as they are faster, give more reliable medical diagnoses and expose patients to lower radiation doses than do current technologies. Today, the commercialization of such advanced detection techniques is impeded by the difficulty of growing large CdTe single crystals with the required quality
- Enhancement of Oil Recovery- Understanding fluid physics in crude oil reservoirs is essential for optimizing exploitation. This is a major challenge because the role of thermodiffusion (the Soret effect) in petroleum reservoirs is not yet fully understood. Understanding diffusion processes is a key element and accurate coefficients of pure diffusion can be measured only in microgravity. As a result, ESA is sponsoring two projects on the precise measurement of diffusion coefficients in crude oil mixtures
- Atomic Clock Ensemble in Space (ACES)- The microgravity environment aboard the ISS allows longer observation times of laser-cooled atoms than is possible on Earth. This translates directly to increased frequency resolution and, hence, more accurate timing. An accuracy of better than 10^{-16} s is expected. This experiment will also provide a variety of applications of a fundamental and technological nature. These include a 30-fold improvement in the measurement of the gravitational red-shift constant, improved testing of the isotropy of the speed of light, a search for a drift in the fine-structure constant, dissemination of time with 30 ps accuracy, a 10^{16} Hz oscillator and also positioning with millimeter precision. ACES has been selected to fly in 2002.

The Weak Equivalence Principle (WEP) Experiment - An experiment will test the WEP for anti-matter to a precision of about one part in a million. It is proposed to store protons and anti-protons in a magnetic bottle with the symmetry axis orthogonal to the Earth's gravitational field. In the experimental set-up, the effect of gravity on the anti-protons will show up as a perturbation to their motion [Reference 57].

The Fluid Science Laboratory (FSL)- Occupying one ISPR, the FSL provides a multi-user research capability for the study of dynamic phenomena in fluid media in the absence of gravitational forces. The most significant element in the FSL is the Facility Core Element which houses the standardized Experiment Containers (ECs), the central experiment module and the optical diagnostics module. The FSL can be operated by the flight crew in automatic or semi-automatic modes or in a remote control mode from the ground (telescience).

The Material Science Laboratory (MSL)- The MSL occupies one ISPR in the Columbus Laboratory and about half of an ISPR in the US Lab. Except for minor rack interface differences, the two MSL versions are identical. A multi-user facility, MSL supports research in crystal growth with semiconductors, the physics of liquid states, measurement of thermophysical properties and solidification physics.

The European Drawer Rack (EDR)- The EDR, occupying part of an ISPR, provides a modular capability for sub-rack payloads. Its fundamental goal is to accommodate smaller payloads in standard experiment drawers (SEDs) and Mid-Deck Lockers (MDLs) for experiments with a quick turnaround.

Japan- Microgravity sciences are promoted by Japan's Space Utilization Promotion Office, with participation from NASDA and MITI. Furthermore, industry interest in microgravity science is developing in Japan. The focus is on the determination of the thermophysical properties of molten semiconductors. This is to satisfy Japan's goal of developing high-quality silicon single crystals for the next generation of miniaturized electronic devices. Other topics of interest include the investigation of technical combustion processes to better understand phenomena such as droplet evaporation and ignition which is expected to result in a reduction of fuel consumption and exhaust emissions.

- Gradient Heating Furnace (GHF) Payload- To conduct research on material sciences, this furnace (GHF), which has three heating zones and a cooling zone
- Advanced Furnace for Microgravity Experiment with X-ray Radiography- This multi-user image furnace (AFEX) consists of a gold-plated ellipsoidal mirror where experiment material is placed at one focus of the reflector and heated and melted by the radiation from a 1500W-halogen lamp. An image of the material can be measured from the other focal point
- Fluid Physics Experiment Facility Payload- This payload is used for conducting fluid physics research on the moderate temperature fluid convection in liquid bridges. With this facility, it will be possible to obtain a 3-D visualization of the flow patterns, resulting from convection, from the laminar to a turbulent state. A 2-D fluid visualization will also be possible from CCD cameras or the infrared imager system. Furthermore, flow velocity measurements on the sample surface and ultrasonic velocimeter readings are available.

NASA- Increasing microgravity science research requires two purpose-built facilities [Reference 58]:

- Fluids and Combustion Facility- The Advanced Fluids Module experiment rack will consist of a number of experiment-specific test chambers which will each carry ancillary equipment such as cameras, laser optics, and heaters to accommodate research in the fields of interface configuration, thermocapillary flow, particle dispersion and gravity-jitter. The Combustion Module will have several viewing ports to allow for various diagnostics as required for research in the topics including comparative soot-flow diagnostics, forced-flow flame spread, fiber-supported droplet combustion, and radiative ignition and transition to spread
- Materials Science Research (MSR) Facility- This facility will support a wide variety of investigations and classes of materials. Furthermore, the design of the MSRs allows them to be operated as autonomous, stand-alone racks [Reference 59].

CSA- The CSA is still engaged in strategic planning for Canadian microgravity experiments on the ISS. The agency intends to have at least one experiment in each of the following areas: (a) fluid physics (molecular motion, bubble motion, interface physics), and (b) materials science (semiconductor crystal growth and diffusion studies in metals) [Reference 60].

Furthermore, the Canadian Space Agency will provide hardware to be used in microgravity research [Reference 61].

- Microgravity Vibration Isolation Mount (MIM)- This mount isolates microgravity experiments from the vibrations inherent to orbiting space platforms. The MIM has been used on Space Shuttle missions and on Mir in the past, and it is intended to be used on the ISS in the future
- Commercial Float Zone Furnace (CFZF)- Created by the CSA and the former German Space Agency (DARA), this furnace melts down sample materials and lets them resolidify in microgravity, allowing new structures to be created. Materials created using the CFZF may be useful in laser devices, medical treatments, fiber optics, and microwave cellular communications. The CFZF was very successful aboard the Space Shuttle, and may be used on the ISS.

RKA- No information was found pertaining to the Russian Space Agency's intentions with regard to the utilization of their laboratories for microgravity research.

Critical Analysis- In spite of the fact that dampers may be used to protect microgravity experiments from moderate perturbations, the microgravity conditions aboard the ISS are less than optimal and even inadequate for certain classes of experiments. In order to carry out some of the experiments described in this section, coordination with astronaut activities that induce vibration (like exercise) may be required. Many microgravity experiments can be automated and accommodated on free flyers in higher orbits where lower drag forces translate to better microgravity conditions.

For onboard space physics studies, active instruments and ground communications interfere with natural radio observations; the neutral and plasma environments are changed by the presence of the ISS to the extent that *in situ* measurements of the natural state are not possible.

5.5 Observing Sciences

Astronomy- The current utilization of the ISS for astronomical research focuses on sky surveys that use equipment which is not too sensitive to stability or pointing accuracy and that are also resistant to pollution from the ISS. There are plans or proposals to do the following astronomical experiments on ISS.

SPOrt (Sky Polarization Observatory)- This payload will survey the sky to measure the polarization of light from all directions. It will increase current knowledge of the linear polarization of the galactic radio emission giving the sensitivities required to detect structures in the Cosmic Microwave Background at levels of a few mK over angular scales from arcminutes to degrees [Reference 62].

UTEF (Ultraviolet Telescope Facility)- This is an ultraviolet telescope that was originally proposed to ESA as a separate telescope but later recommended as an ISS telescope. A hexapod system would be used to achieve the required pointing accuracy and attitude control, and the telescope would use active optics to achieve the desired stability.

AMS (Alpha Magnetic Spectrometer)- This instrument is a large magnet with detectors that will study cosmic ray propagation in the galaxy. It requires the space environment to search for anti-matter and dark matter in space [Reference 63].

X to Gamma Ray Sky Survey- This survey will use a coded mask telescope to observe the sky in the hard X-ray to gamma ray range. It would combine a wide field of view with resolution high enough to achieve accurate source location.

OPAL, Light Scattering Facility- This instrument will be used to make optical measurements of dust. It would use the microgravity conditions on board ISS to simulate interstellar or interplanetary dust and observe how it interacts with light. These results could then be used to better understand the effects which such dust has on our observations of stellar and solar light.

Earth Observations- A wide range of remote sensing observations is possible.

ESA- Current and planned operational space-borne Earth observation systems provide spatially and radiometrically crude data for the detection and monitoring of high temperature phenomena on the surface of our planet. High temperatures often result from environmental disasters such as forest fires, Savannah fires, and volcanic activity. For this reason and others, a global environmental monitoring system from space is required.

• Fire Detection Infrared Sensor System (FOCUS)- Based on pilot studies and experimental work, the ISS will be used in its early utilization phase as a platform and test bed for an intelligent infrared sensor prototype, FOCUS, a future Environmental Disaster Recognition Satellite System (EDRSS). FOCUS is considered an important mission combining a number of proven technologies and observation techniques to provide the scientific and operational user community with data for classification and monitoring of forest fires [Reference 64]
• ALADIN- Within the framework of the Earth Explorer Program of ESA, the Atmospheric Dynamics Mission (ADM) is planning to provide wind field measurements on a nearly global scale. The Atmospheric Laser Doppler Instrument (ALADIN) will be accommodated on the ISS and will be used to investigate the diurnal cycles over three years at latitudes covered by the Space Station. Climatological and meteorological modeling and forecasting will be improved by these global wind field measurements [Reference 65]
• Real Time Mapping using Multispectral Spaceborne Videography- A real-time space-borne (ISS) multispectral videography system would provide compressed video data together with orbital data back down to Earth using the existing down links to a central server. However, this idea is based on the fact that a Global Positioning System (GPS) would be available on the ISS, but plans for the onboard GPS have been altered. Therefore, this is quite a doubtful utilization idea as it stands now [Reference 66].

NASA- The Office of Earth Science (OES) of NASA will make use of the ISS as one of the platforms for Earth observations. The one payload currently planned for flight on the ISS in 2002 is the Stratospheric Aerosol and Gas Experiment (SAGE III). Also, NASA is considering the KidSat Science Educational Payload to be on ISS [Reference 67].

- SAGE III- This instrument will continue observations of the vertical distributions of ozone, aerosol and other trace constituents pioneered through SAGE I and SAGE II. SAGE III has the advantage of both solar and lunar occultation to measure aerosol and gaseous constituents of the atmosphere during Space Station sunrise and sunset events
- KidSat- The KidSat video camera will be accommodated on the ISS for observations of clouds, land, ocean and other phenomena, and transmitting data to schools in real time on the Internet. Teachers and students will be engaged in a collaborative learning process of conducting scientific experiments in space using NASA resources and the national educational infrastructure. This should capture the interest of all children and help educators motivate children to excel in science, mathematics and engineering.

NASDA- In the Japanese module, there will be a Superconducting Submillimeter – Wave Limb Emission Sounder (SMILES), a highly sensitive submillimeter wave limb sounder using a superconducting technique which enables the three-dimensional global observation of trace gases in the stratosphere, of relevance to ozone depletion and global warming [Reference 68].

RSA- No information on Russian plans for Earth observations could be found.

CSA- Canada has no plans for Earth observations from the ISS.

Critical analysis- Although the orbit of the ISS (51.6° inclination) provides an Earth surface coverage of 85%, it is, however, not optimum. With this orbit it is, for example, not straightforward to study important phenomena like ozone depletion, occurring over the poles. Moreover, due to non Sun-synchronous orbit, lighting conditions on the Earth's surface vary for repeated tracks. Also, the periodic variation of the orbital height of the ISS might pose some limitations for Earth observation applications as it directly translates to a variation in the resolution of instruments.

Another limiting factor of the ISS is that vented gases and light pollution contaminate the optical observing environment. Solar panels and other installations should not obscure instruments' fields of view. Many of the important factors related to doing astronomy from space, such as orbit, attitude control, stability, and communications, are not optimized for the individual payloads. This places several restrictions on the use of ISS for astronomical purposes:

- LEO- "Perhaps the worst orbits are those offered by 'infrastructure', which were not designed with astronomy as their primary driver yet have to be used 'because they are there'" [Reference 69]. The low Earth orbit is generally not suited for most astronomical purposes as the Earth frequently blocks its view or scatters light into the instrument. Its passage through the Van Allen radiation belts can also interfere with observations. This orbit is sufficient, however, for some missions such as all-sky surveys
- Stability and Attitude Control- Many astronomical observations need additional stability and attitude control systems to achieve the required pointing accuracy. However, there are some instruments that do not require precise stability and attitude control, such as coded mask telescopes. And there are devices, like the hexapod base, which can compensate for these problems
- Environment- The local environment will be polluted by water vapor and waste gases, eliminating the possibility for cryogenically cooled instruments.

The ISS could be useful, possibly, for testing new kinds of space-based astronomy instruments which have not yet been space qualified or need frequent adjustment to get them working in space. Most astronomy, space physics and Earth observation investigations can be more cost-effectively and successfully conducted from unattended satellites [Reference 70].

6. How to Utilize the International Space Station

6.1 ISS Management and Planning

The ISS planning process is very complex and involves many interfaces and products. It is also being performed in several distinct steps with distinct planning processes covering different time intervals ranging from several years to just a few days. As a consequence of its unique multinational character, the planning for the ISS is based on a distributed hierarchical concept. Each Partner in the Space Station Program is considered a single planning entity, and, distributed geographically, each is tasked to integrate and plan their respective Partner element activities and resource utilization [Reference 71]. To facilitate understanding of the planning process, two terms must be introduced and explained [Reference 72]:

- Increment (I). This is the time from the launch of a manned vehicle to the undocking of the return vehicle for that crew. The length of an increment ranges anywhere from 1 month to about 6 months. This term refers to all of the activities occurring during this time frame

- Planning Period (PP). This is the period on which much of the ISS planning is based. It spans approximately 1 calendar year, but is linked to the beginning and end of ISS increments, and so usually does not begin on January 1.

Strategic Planning- Each Strategic and Multi-Increment plan covers a five-year period through scrolling plans that are updated regularly. Each year the new period "P5 " is added and the period of the previous year "P1" is removed. Details for years "P4" and "P3" are refined while the data for "P2" and "P1", those closest to execution, will then reflect the information provided by the tactical planners. The Multilateral Coordination Board (MCB), which comprises representatives from each of the Partner's agencies, has established Panels responsible for the long-range strategic coordination of the operation and utilization of the ISS. They are the System Operations Panel (SOP) and the User Operations Panel (UOP) [Reference 73]. These two panels will develop the annual Composite Operations Plan and the annual Composite Utilization Plan, respectively. The latter is the consolidation of the various Partners' utilization plans. Based on these two documents, SOP and UOP jointly prepare the Consolidated Operations and Utilization Plan (COUP) which is the end product of the annual strategic planning cycles.

A further multilateral body, the Multi-Increment Planning Integrated Product Team (MIPIPT), develops the Multi-Increment Manifest. This document defines the traffic and crew rotation plans for the five planning periods contained in the COUP.

Tactical Planning- Tactical Planning starts 30 months in advance of the Planning Period. This covers approximately one year, generally from the first manned launch in the calendar year until the first launch of the next calendar year. This planning period will usually entail four increments. Based on the COUP requirements, the international Partners develop their proposal according to their flight-element and user needs. The main product of this planning process is the Increment Definition and Requirements Document (IDRD), signed by all the international Partners. The preliminary version of the IDRD is released at Planning Period Start (PPS) –24 months and the baseline is published at PPS-18 months. Further updates take place every 6 months, as required. The IDRD serves two purposes [Reference 71]:

- For each specific increment mission, it defines operations and utilization objectives, top-level cargo manifest (up-/down-loads), payload complement, accommodation and resource allocations, crew rotation and training plan and in-orbit maintenance

- In addition, a set of summary documents is provided which are used to inform the strategic-level planners about what has been planned for P1 and P2 of the upcoming COUP and the degree to which these IDRD's have met the previous COUP's requirements.

Other planning documents at the tactical level are the Engineering Feasibility Assessment (EFA) and the Operations Feasibility Assessment (OFA). They are meant to ensure that the priorities and objectives of the increment can be achieved.

Increment Planning- Planning at this level can be divided into the pre-increment planning phase and planning during the increment. An International Execute Planning Team (IEPT) is tasked to perform the above planning based on the IDRD as the major input. The most important outputs for the phases mentioned above are the On-Orbit Operations Summary (OOS) and the Short-Term Plans (STP), respectively.

The OOS utilizes a resource profile analysis, carried out on a weekly basis, which defines the possible limits for each resource (crew time, energy, communications, etc.) over a defined period of time, usually the increment itself. The OOS can then be developed (PP-18 months). The Preliminary OOS is delivered at PP-12 months with the final version published at PP-2 months. It will be updated through to the end of the planning period to reflect operations as they actually occurred.

The STP is a detailed, integrated schedule of activities to be performed during one week of Space Station operations. The STP includes all ISS operations, including each of the ISS Partner systems and payload activities. Also derived from the STP is the On-board STP, which is the integrated plan that is executed onboard the ISS. It will contain three days of activities (yesterday's, today's and tomorrow's).

6.2 Payload Cycle and Integration

The payloads onboard the ISS are divided into two categories:

- Class 1, or Facility Class, Payloads can be regarded as permanent multiple-user facilities (complete rack facilities or complex instruments on an Express Pallet Adapter) that provide services and accommodation for experiments
- Class 2 payloads are those that are embedded into a facility and are provided directly by the users (scientific and/or industrial users).

The selection and procurement processes are different from one payload class to the other. Class 1 Facilities are generally developed at the Agency level in co-ordination with the other ISS Partners; Announcements of Opportunity are issued for the Class 2 payloads during the development of the Class payloads and throughout their exploitation phase. The Russian Space Agency (RKA), however, has requested proposals from the scientific communities prior to the development of their "Class 1" facilities for ISS.

The main difference in the development process between the two classes of payloads is the time needed to complete the various stages. Based on the experience acquired with Spacelab, the development time for a Class 1 payload can last up to 5 years while, in comparison, the time needed to develop a Class 2 payload can vary from a few months to up to three years.

Following successful transition through the ISS partner's selection process, the payload is included in its PUP (Partner Utilization Plan) which is made on an annual basis and five years in advance of utilization. The Plans are forwarded to the UOP that will develop the CUP. A PUP proposed by any ISS Partner that falls completely within their respective allocations and does not conflict with another Partner's Utilization Plan will be automatically approved [Reference 74].

The payload (P/L) development cycle is summarized in Table 5 [Reference 75]:

Payload Activities	Supporting Activities
Conceptual Design	Strategic Planning
Feasibility Study (Phase A)	
	Tactical Planning
Design Phase (Phase B)	
Development	Interface Agreement
Qualification	Analytical Integration
Verification	Operations Preparation
System Compatibility Testing	Final Acceptance / Documentation
System Integration	Physical Integration
Launch Preparation and Launch	Logistics Support/Final Checkout
On-Orbit Verification	
On-Orbit Operation	Flight Operations
Return from Orbit	

Table 5. Payload development phases

Payload (P/L) Integration is carried out at the Payload Operations and Integration Center (POIC) located at the Marshall Space Flight Center (MSFC) in Huntsville, Alabama, USA. The POIC will fulfill a double role. At the broad level, in co-operation with the Space Station Control Centers (SSCCs), it will act as Space Station Integrator, receiving resource requests from the element planners (COF, US Lab, etc.), working out resource profiles for P/L operations and assigning them to each P/L operation. At the Payload Planning Center level, it will look after all the US payloads either located in the U.S. Lab or in any other element of the ISS (e.g., US payloads in COF).

6.3 How to Gain Access to the Space Station

A number of experiments can be performed during the assembly phase when the Space Shuttle and Russian launchers visit the Space Station and between flights when the on-board crew is available as experiment operators or as research subjects. A set of research hardware will be transported to the Space Station, primarily in dedicated Space Shuttle Utilization Flights (UFs). So far only seven of these flights have been scheduled, starting with UF1 in 2000 through UF7 in 2003. From this point onward, Shuttle and additional logistic flights from the Partners' mixed fleet are planned each year for exchanging the astronauts and re-supplying the payload items.

Scientific Investigations- Through Announcements of Opportunity, the International Partners are already selecting scientific experiments for the ISS. The proposals undergo an evaluation and selection process through a peer-review system. The evaluation criteria applied by the different Partners are basically the same:

- Scientific/technical relevance and merits
- Assessment of the technical feasibility of the proposed experiment
- Appraisal of the proposal relevance to the program priorities of the sponsoring Partner Agency
- Qualification of the team
- Quality of the preparatory research on which the proposal is based
- Probability of achieving the scientific/technical objectives.

It is noteworthy that, in the effort to develop cooperation, the Space Agencies and then participants are meeting in an International Microgravity Strategic Planning Group (IMSPG). One of the objectives is to coordinate the development and use of research apparatus among microgravity research programs in areas of common interest to maximize the productivity of activity internationally. The group is presently discussing ways and means to intensify international co-operation in the context of utilization of the ISS. A first International Announcement of Opportunity in the field of Life Science Research (LSRA) was issued in December 1996 [Reference 76] and another is planned for 2000/2001. Internationally coordinated AO's are being prepared also for the disciplines of Physical Sciences and Biotechnology.

6.4 NASA's Approach to Exploitation of the ISS

Commercial Development and the Commercial Space Centers (CSC)-NASA is strongly committed to explore commercial exploitation of the ISS facilities. To this end, NASA has already committed 30% of the ISS's payload capacity for commercial development. Internal studies are being conducted to identify potential areas of commercial opportunity [Reference 78]. In addition, a Commercial Development Plan for the ISS [Reference 79] has been prepared. Its goals are, in the short term, to begin the transition to private investment and offset a share of the public cost, and, in the long term, to establish private sector demand for space products and services in LEO.

Among the variety of methods used by NASA to solicit and select commercial projects for flight onboard the ISS, the best established is NASA's designated Commercial Space Center (CSC) [Reference 79]. These Centers are a joint undertaking funded through NASA's Office for Life & Microgravity Science and Applications (OLMSA). They involve teams of industry, university and other non-NASA government organizations which were formed to provide a pathway for the US to develop new products, processes, markets and services. Each of the Centers has industry commitments in the form of cash and in-kind resources to carry out the programs. When a party has a proposal that is of potential economic benefit, it can approach a CSC with a proposal. This partnership generally begins with "expert-to-expert" discussion and progresses to an industrial commitment, the development of a research plan and its implementation.

In the frame of NASA's commercial development program, the CSC's have been tasked to query their existing industrial affiliates and evaluate the prospect of the Space Station becoming a fee-for-service product development laboratory or production center.

Table 6 presents a list of the current CSCs (approved or in process of obtaining NASA's approval):

CSC Name	Location	Field of Activity
BioServe Space Technology	University of Colorado, Boulder, Colorado	- Bioprocessing/bioproduct development - Bio-molecular electronics - Physiological modeling
Center for Commercial Applications of Combustion in Space	Colorado School of Mines, Golden, Colorado	- Combustors, fire safety, ceramic powder
Center for Commercial Development of Space Power and Advanced Electronics	Auburn University, Auburn, Alabama	- Thermophysical properties of casting alloys
Center for Macromolecular Crystallography	University of Alabama, Birmingham, Alabama	- Three-dimensional structure of protein crystals
Consortium for Materials Development in Space	University of Alabama, Huntsville, Alamba	- Space materials
NASA Langley	Hampton, Virginia	- Polymer materials
Microgravity Automation Technology Center	Environmental Research Institute of Michigan, Ann Arbor, Michigan	- Laboratory automation
Space Vacuum Epitaxy Center	University of Houston, Texas	- Ultra-pure, ultra-thin materials for electronics
Center for Space Automation and Robotics	University of Wisconsin-Madison, Wisconsin	- Technologies for long duration space missions
Worcester Polytechnic Institute	Worcester, Massachusetts	- Crystal growth

Table 6. NASA's Commercial Space Centers (CSCs)

New Concepts in ISS Utilization Management- NASA is aware that in order to make the ISS an effective and successful program, management must ensure its productivity and optimization of the results by the R&D and business communities. In order to achieve this, the establishment of a non-government organization (NGO) has been proposed in the United States by the year 2000 [Reference 80]. The purposes of this new organization (see Table 7) are to coordinate the science and the engineering communities to "aggressively expand the scientific (...) and technological capability of the ISS" as well as "disperse information on the resulting scientific and technological achievements", and to stimulate "the commercial community to expand the global economy in space products and services". As far as customers ("sponsors") are concerned the main distinction is made between public sponsors and private ones. The latter enjoy Proprietary Rights on the R&D results, that is, the results will not be disclosed and will remain the property of the funding sponsor.

Office of the Director - Selected by the Board of Directors - Utilization program development - Management and administration	Operations Board - Select projects' scientists & engineers for residency - Approves visiting senior scientists & engineers - Assigns Mission Directors and R&D Working Groups - Approves payload integration plans & flight assignment - Assigns operating periods & accommodation sites
Board of Directors - Annually reviews & extends research programs - Communicates policies of the sponsoring organizations - Approves Annual R&D Program Plan and Commercial Prospectus	Science Program Office - Scientific research program management - Conducts nominal share of scientific research - Establishes science project queue - Defines requirements for flight instruments - Manages analytical, physical and operations integration - Manages science results archive
Liaison Office - Staffed by national and international program sponsors - Represents sponsors and provides oversight	Technology Program Office - Technology development program management - Establishes technology project queue - Defines requirements for flight equipment - Procures/develops flight equipment - Manages analytical, physical and operations integration - Manages technological results archive
Education Office - Develops collateral products for education - Communicates attributes of orbital environment and achievements of the R&D programs	Bonded Commercial Program Office - Implements commercial policy of government sponsors - Liaises with private sector and commercial space network - Establishes commercial project queue - Manages analytical, physical and operations integration - Maintains proprietary procedures and protocols
Operations Office - Strategic, tactical and contingency planning - Manages resource allocation and mission model - Manages mission support contract - Produces annual R&D Program Plan and annual Commercial Prospectus	Space Trust Corporation - Manages private capital funds - Selects private ventures for funding with equivalent rigor to private capital markets - Finances qualified private ventures, if necessary

Table 7. Proposed structure and duties of the NGO

6.5 The European Approach to Exploitation of the ISS

Access to the ISS for European undertakings is mainly organized and managed by the European Space Agency. In addition to the overall European payload, nationally provided payloads must also be taken into account. Unlike NASA, Europe has devoted its attention mainly to scientific and technological utilization of the ISS. Access and charging policies addressing industrial/commercial users are still under development.

Bearing in mind the ISS planning process, it is important that any Partner be able to make a long-term commitment toward the operation and utilization of the Space Station. To this aim, ESA is currently preparing an overall Exploitation Program for the period 2000-2013, as an optional program.

Even though firm rules for user access are still under discussion at ESA, users can be defined in three main categories: Fundamental and Applied researchers from ESA Member States participating in the ISS exploitation program; Industrial/Commercial Users from Program participating Member States; and so called "Third party Users", which are all the undertakings of ESA Member States that are not participating in the ISS and/or non-Member States.

A set of basic procedures will be applicable to all users: the AO process, a technical feasibility assessment by ESA, and review and recommendation by the ESA Program Board concerned (e.g. Microgravity, Space Science, etc.). The main difference in the ways in which different user categories will be treated lies in the charging policy. Third Party Users will, in most cases, bear the full mission cost or have to pay an access fee fixed by the ESA Executive on a case by case basis. It should also be possible for a National Space Agency to select, develop and fly an instrument and nominate the user for it according to its own criteria and outside the ESA selection loop. In this case, the cost will be borne by the proposing Agency.

Even though the full operation of the Columbus facility will occur only in 2003, through a dedicated MoU, NASA has arranged to provide ESA with early utilization opportunities [Reference 81]. In exchange for ESA's provision of so-called "Space Station utilization enhancement items" (the Microgravity Glovebox, three -80 Degree Celsius Freezers, the Hexapod Pointing System and the Mission Data Base system) NASA will provide ESA with the following: access to 50% of a NASA payload rack in the US Lab for two years; the equivalent of up to four mid-deck lockers to accommodate European-provided research facilities; the use of one-half of an attached payload accommodation site on the truss and payload participation on two Shuttle flights.

User Support Operations Centers- for payload flight operations management, Europe has opted for a decentralized approach. The focal point will be the Columbus Operations Control Center (COF-CC). The COF-CC has been assigned an overall monitoring and coordination task at the element level of the different Microgravity Facilities for Columbus (MFC), such as Biolab, FSL, MSL, EPM and EDR. Preparation and in-flight operations for the MFC during each mission increment have been assigned to User Support Centers under the overall direction of the COF-CC.

These centers are the link between the payload operations onboard the Space Station and the scientific user group. They are involved to a different extent in the preparation and mission execution function. A center appointed as a Facility Responsible Center (FRC) prepares and operates the facility assigned to it. Currently, the following European centers have been identified as possible FRC's:

- Microgravity User Support Center (MUSC) in Cologne (EPM, MSL-SQF)
- Microgravity Advanced Research and Support Center (MARS) in Naples (FSL)
- Centre d'Aide au Developpment de la Micropesanteur et aux Operations Spatiales (CADMOS) in Toulouse (Biolab and MSL-LGF).

6.6 The Canadian Approach to Exploitation of the ISS

To help Canadians utilize Canada's share of the ISS, the Canadian Space Agency has formed the Microgravity Sciences Program (MSP). It assists scientists in taking advantage of the microgravity environment aboard the ISS to advance research. The MSP's Announcements of Opportunity (AO) are published to solicit proposals for experiments from Canadian researchers. These teams may include international scientists.

Canada has also submitted a Partner Utilization Plan (PUP) that indicates it will start utilizing the ISS in 2001. The CSA is currently evaluating the ISS as a test-bed for on-orbit technologies. It may commercialize part of its ISS utilization share since it will not make use of all the resources available over the lifetime of the program [Reference 82].

6.7 The Japanese Approach to Exploitation of the ISS

In addition to participating in the International Announcement of Opportunity for life science experiments on ISS, Japan has its own module (JEM) and a plan for utilizing it. There are two types of facilities on JEM, the Pressurized Module (PM) and the Exposed Module (EM). Proposed experiments are selected through the AO process and there are currently fifty proposals in the PM and four in the EM. After experiments are selected, common facilities are developed for them and the next generation proposals are selected for these facilities.

7. New Utilization Opportunities, and Recommendations for the Future

A thorough examination of the literature on the International Space Station suggests a variety of possible utilization schemes. Although the possibilities are boundless, this overview examines future utilization in terms of current planned usage which is predominantly scientific experimentation. There are three areas of scientific research discussed: life sciences, physical sciences and observing sciences.

There is a wealth of potential for life science research onboard the ISS. Space medicine in particular stands to be advanced significantly using the Russian tradition of studies on long duration missions as a base. There is much more significant work to be done in preparation for mass habitation and travel in the unforgiving outer space environment.

7.1 Life Sciences

Developmental Biology- Research in post-natal mammalian development should be a key area of research on the ISS. The elucidation of mammalian development is a preliminary stage to understanding human developmental processes. As mentioned previously, post-natal studies using rodents are hampered in microgravity due to poor maternal care and feeding of the offspring. If appropriate research is performed on this problem aboard the ISS, it is expected to yield key results, and bring in rich dividends for biomedical research.

One of the main problems with studying the effects of microgravity on plant biology has been the lack of long duration spaceflights. The ISS provides a platform for long duration microgravity and should therefore be exploited for research into plant biology. Understanding the behavior of plants in microgravity is necessary to develop bio-regenerative life support systems and to understand fundamental processes applicable to agriculture on Earth. An area of plant research that has particular importance is cultivation through multiple generations. Such a process may be required in a bio-regenerative life support system and should build upon the successful cultivation techniques used on previous missions aboard Salyut and Mir.

Tissue culture- In microgravity, cell cultures may develop complex three-dimensional structures that closely resemble certain tissue structures. This "tissue engineering" has great potential in the area of organ transplantation.

Protein Crystallization- The microgravity environment enables the production of highly regular crystal structures from protein solutions. These structures can be analyzed on Earth using X-ray diffraction or nuclear diffraction. The complete structures of many human proteins have not yet been elucidated. Given the immense benefit to the medical field offered by these studies, full advantage should be taken of the opportunities on ISS for protein crystallization studies.

Radiation- The long term effects of space radiation on plants, animals and the human body are unknown. Although tumor induction, life shortening, cell transformation, and chromosomal mutations and translocations have been observed in some animal studies, other factors, such as stress and genetic background, may be involved. The ISS will provide a unique facility to study long term exposure to radiation.

Space Medicine- In the field of space medicine, much research can be effectively carried out aboard ISS.

Skeletal Muscle- Muscle atrophy (specifically loss of gravity-resistant Type 1 "slow" muscle fibres) is a significant problem for astronauts following long-term spaceflight. Treadmill and ergometric exercises are not effective at preventing this. To assist re-adaptation of ISS astronauts to 1g following their return to Earth, an effective counter-measure must be developed on the ISS that prevents the loss of "slow fibres". Such a counter-measure may improve upon previous "constant-loading" suits (e.g. the "penguin-suit") worn by cosmonauts onboard Salyut and Mir. Such suits may assist the treatment of conditions experienced on Earth, such as cerebral palsy in children.

Bone- The bone-weakening symptoms experienced by cosmonauts on long-term space flights are similar to those experienced by osteoporosis patients on Earth. Development of an effective counter-measure to bone weakening in astronauts may therefore benefit medical treatments on Earth. Such a counter-measure may require imparting shocks/vibrations to the gravity-resisting bones (e.g. femur) to induce osteoblast (bone-forming cell) activity. This measure could obviously disturb other microgravity experiments on ISS. To compensate, it is suggested that an important criterion for the assessment of ISS experimental proposals is the ability of an experiment to generate scientifically useful data and to retain all previous data during and following disturbances to the microgravity environment. That is to say, experiments on the ISS should be relatively "resistant" to microgravity disturbances.

Cardiovascular System- Important, but so far little studied, areas of the cardiovascular system are 1) coronary and 2) cerebral blood flow. Human studies on the ISS should include ground-based measurements from subjects not only in supine positions, but also in upright positions (with relative central hypo-volemia). This is essential for scientifically valid comparisons with measurements taken during the later stages of the mission when total blood volume is decreased.

Nervous System- Measurement of inner eye pressure in space using a 'self-tonometer' has applications on Earth. The same is true for video-oculography; it can be used in the detection of disturbances in the vestibular, neurological or oculomotor domains. A refined version of the opthalmo-dynamometer could be used to measure raised intra-cranial pressure in astronauts which may be linked to space motion sickness. The investigation of this non-invasive method may benefit patients on Earth that require neuro-surgery. However, careful interpretation is required positively to correlate ocular pressure with intra-cranial pressure.

Respiratory System- Closer investigations into the effects of long term microgravity on the mechanics of breathing and on the neuro-physiological adaption of the respiratory system (with changed bio-mechanical conditions) are needed. In addition, a sensitive respiratory sensing device should be developed for testing physical fitness in space (and assessing cardiovascular de-conditioning).

Psychology-Previous reports demonstrate that social behaviour and mood changes are significant factors that affect the success of a space mission. This area of research should be given great importance on the ISS, considering the international nature of ISS crews, and their cultural diversity. In addition, crews should be given sufficient pre-flight time to work together as an effective team.

Medical Care in Space- The ISS can be used as a test-bed for ambulatory monitoring devices, biomedical instrumentation, tele-medicine systems, life support, and environmental control systems. These systems will have direct benefits to people on Earth.

7.2 Physical Sciences

Observation of Auroras-Human observations of such physical phenomena are particularly useful because real-time intelligent judgements of what is interesting and what is related to what can be made. Although the ISS orbit is not highly inclined, oblique observations of auroras in the polar regions will be possible. During particular periods of interest such as geomagnetic storms or in support of ground-based campaigns, astronauts could be asked to observe and film interesting auroral phenomena in response to specific requests from interested scientists.

Sprites-These are lightning flashes, which originate at the tops of clouds and strike upward towards the ionosphere. They have been reported in the scientific literature only this decade and the mechanism for their generation is not yet fully understood. One of the factors hindering a fuller understanding is a lack of geo- and time-referenced optical data of these phenomena. Several features of the ISS make it an ideal platform from which to record sprites. Sprites typically occur over mesoscale convective systems (MCS), i.e. large thunderstorms. These conditions can be ascertained using existing ground- and space-based meteorological observation networks. Once an observation opportunity is identified and if an astronaut is available during such a period, he or she could use the cupola, if it were provided with an accurate pointing system, or an external camera, to record these observations. Finally, the program could also make use of ISS's long life by studying the effects of the eleven-year solar cycle on the occurrence of sprites.

Quantum Computers- With the development of more precise atomic clocks on board ISS comes the opportunity to develop more powerful quantum computers. These computers are currently being developed on Earth and will be exponentially faster than conventional computers. The microgravity environment, which improves the accuracy of atomic clocks, might increase their computing power. With the development of these computers comes the opportunity to study the quantum principles that govern them, giving us an excellent laboratory on board the ISS.

7.3 Observing Sciences

In-Orbit Assembly- The ISS could be used to assemble very large structures. One obvious candidate for in-orbit assembly is a large, multi-aperture interferometer for astronomical observations, while removing the need for complex automatic assembly systems. Since the separation of these mirrors must be maintained to within a fraction of the wavelength of light, astronauts would be needed to make adjustments from the beginning.

Earth Environment Observations-NASA proposes payloads to research:

- Land cover and land use
- Climate change phenomena
- Atmospheric dynamics.

In such experiments, which should complement other satellite studies, information on the nature and extent of environmental conditions and consequences for sustained development in the tropics and mid-latitude regions should be obtained. Improved forecasts of precipitation and temperatures, on seasonal to inter-annual timeframes, to predict natural hazards and to mitigate natural disasters in the tropics and mid-latitudes should be obtained.

8. The Future

Now that future uses of the ISS for its original purpose, scientific experimentation, have been examined, it is time to consider some other ways to use the International Space Station that may not have been considered in the design. As a base for expanded operations in outer space, the ISS has great potential. Three possible activities touched upon in this overview merit special consideration for utilization: 1) as a base for other free-flying platforms, 2) as a jumping-off point for resource recovery and future missions of exploration, and 3) as a construction platform.

Other free-flying platforms will soon be needed in low Earth orbit. Microgravity labs are a possible first application. Throughout this paper, the unstable microgravity environment on the ISS has been described. Scientific users doing microgravity research seem more than willing to adjust their objectives to compensate; the quality of the science suffers. Certain areas of microgravity research like fluid dynamics are not worthwhile under these conditions. Because compelling microgravity research will certainly be one driver of Man's ventures beyond Earth, free-flying labs are necessary to ensure high quality experimentation. Additional free-flying platforms supported by the ISS might be engaged in manufacturing, satellite salvage and repair, or tourism.

The ISS will likely be used as a staging point for a variety of future missions. Mining operations on the Moon and near Earth asteroids would provide immediate benefits. Because launch costs remain high, any raw material gathered in outer space is inherently cost effective. For the same reason, other missions originating from the Space Station instead of from Earth at the bottom of its gravity well [Reference 51] save precious resources.

Although the ISS has not been designed as a construction platform and EVA resources are severely limited, this application is suggested by a variety of potential activities. Specifically mentioned in this overview is the construction of the next generation of large interferometric telescopes which will be too large to construct on Earth. Other possible construction projects include habitats, solar power and other satellites, and laboratory and manufacturing facilities.

For extensive construction on-orbit to be feasible, however, several advances in thought and practice are necessary. First is the judicious use of pre-launch construction to offset EVA time. Second is the reduction of launch costs. Next is greater EVA efficiency through improved equipment and robotics. Last is the use of extra-terrestrial resources for the construction. In the light of current improvements in each of these areas, it makes sense to retrofit the ISS in ways that would serve the construction application.

No business analysis has been included in this discussion of the International Space Station. It has been a struggle to integrate economics into an otherwise multidisciplinary document, born in a moment of shifting values and expectations of the ISS. As commercialization takes hold, running the ISS in a more business-like fashion will be vital. Certain broad statements can be made, however. The International Space Station has been incredibly expensive to build. It will be incredibly expensive to maintain. Sound marketing of its unique benefits to the user must be presented to make profit a possibility. The overwhelming expense of the ISS suggests that utilization for science only will never pay the bills. But as a foothold for the expansion of mankind into outer space, it will be remembered as a bargain indeed.

While commercialization has long been a favorite buzzword, it is only in recent months that the national space agencies have mobilized actual efforts to approach potential industrial users. Perhaps passage of the Commercial Space Act of 1998 in the USA has energized this. Still, no plan has been advanced that makes industrial profitablity attainable for Earth-bound businesses. The effort to package the qualities of the ISS in an attractive, marketable way is just beginning. Any of the Partners might consider developing a turn-key orbital manufacturing operation based on the specifications provided by the firm or firms that eventually assume ownership. By handling the development costs, one or more governments can ensure the success of the first space industry. The prospective plant should be so profitable that others want to get on the bandwagon of space commercialization. Only profit can drive Man's exploration into space in this time of reduced national budgets.

Hand in hand with a more commercial approach for the ISS should go a feature that has been a struggle from the earliest conceptual stages of the ISS.

And that is comprehensive planning. It is amazing to discover, as more is learnt about the ISS, how incredible the interlocking problems become when planning is short-sighted (or non-existent). Although the COUP is a sound planning framework, its implementation raises as many questions as it answers. Is there suitable research to occupy the ISS after the early utilization phases? Is a long term overview plan, including commercialization, envisioned? Can the Space Station be run profitably by private interests? Have end of life considerations for the project been planned? Because the ISS has a projected life of 10-15 years and Mars missions are scheduled to begin in that time frame, is the ISS destined to twist Mir-like in the political wind in spite of its longer potential utilization? Now that parts of the ISS are actually on-orbit, a comprehensive planning scheme is needed which integrates the full range of economic, legal-political and human issues in with the engineering and scienc issues.

One fact evident from the very beginning of this study is that there are really no new ideas concerning the International Space Station. A lot of the suggestions in the literature have been around for years. Perhaps an integrated, multi-disciplinary planning approach is the new idea. Only comprehensive planning can ensure that the ISS is maximized for the benefit of all.

It is not difficult, living on the edge of the new millennium, to view Man's first steps into space in the way in which we look at the first of our remote ancestors crawling from the sea or the apes coming out of the trees to walk the savannah. We are witnessing and driving the evolution of Man. In this regard, the International Space Station is a foothold, a beachhead on the way to space. With a sound plan and full utilization of the ISS, Man can break out from the beachhead of low Earth orbit and reach the high frontier of human destiny.

Acknowledgements
As always, any edited work with multiple contributors requires a great deal of collaboration. The quick response by everyone involved is greatly appreciated. Several of our colleagues deserve special mention. Jean Pierre Bombled of Arianespace, Alain Gonfalone, head of ESA's Experiments Support Section, Oleg Atkov and Yoshinori Fujimori of ISU all gave particularly thoughtful reviews of the first draft. Ram Jakhu mother-henned the legal section marvellously. David Sylvester and Martha Milkeraitis gave yeoman service as proofing editors in the earliest stages. The original six editors of Literature Review 1 in the fall of 1998 gave liberally of their time and sanity. Jake Maule, Simone Garneau, Rob Alexander, Vaios Lappas, Karin Remeikis and Suparna Madhu will be forever changed by the experience. Dr. Michael Rycroft, ISU's editorial godfather, has provided a calming influence even at a distance. And Nikolai Tolyarenko, Director of the MSS Program, once again demonstrated a knowledge of ISS's technical detail that is truly astonishing. No closure to ISU's Fourth Annual Symposium would be possible without expressing thanks to Patrick French, the first representative of ISU to arrive in France.

References
Section Two-Historical and Political Development
1. Logsdon, J.M.: Together in Orbit: *The origins of the International Space Station*, Monographs in Aerospace History, NASA, 1998
2. Gibbs, G., and Poirier, A.: *Canada and the International Space Station Program: Overview and Status*, International Astronautical Federation Congress, Paper IAF-98-T.1.03, Melbourne, 1-7 October, 1998, American Institute of Aeronautics and Astronautics, 1998

Section Three-The Law Applicable to Utilization
3. European Space Agency: *The Space Station Co-operation Framework*, Bulletin 94, p. 53, May 1998
4. Article 16.3 sub a of the Intergovernmental Agreement of 1998
5. Article 16.2 sub f of the Intergovernmental Agreement of 1998
6. Schwetje, F.K.: *The Legal Regime of the U.S. Space Station*, Proceedings of the 31st Colloquium on the International Law of Outer Space (Bangalore), AIAA, pp. 179-182, 1988
7. DeSaussure, H.: *Tort Jurisdiction Over the International Space Station*, Proceedings of the 32nd Colloquium on the International Law of Outer Space (Torremolinos - Malaga), AIAA, p. 305, 1989
8. Articles VI and VIII of the Outer Space Treaty of 1967
9. Smith, B.L. and Mazzoli, E.: *Problems and Realities in Applying the Provisions of the Outer Space Treaty to Intellectual Property Issues*, Proceedings of the 40th Colloquium on the International Law of Outer Space (Turin), AIAA, pp. 170-171, 1997
10. See note 1 above
11. United States Senate: *Report on Extraterritorial Application of Patent Laws* S 459, November 22, 1990
12. Lecture delivered by Andre Farand.: International Space University, *International Space Station: Law Applicable to Utilization Activities*, Strasbourg, France, 1998

Section Four- Technical Description of ISS
13. NASA Web Site- www.nasa.gov
14. Lecture delivered by Professor Nikolai Tolyarenko: International Space University, *Lectures on the International Space Station*, Strasbourg, France, 1998
15. National Aeronautics and Space Administration (Mission Operations Directorate Space Flight Training Division): *International Space Station Familiarization*, December, 1997
16. European Space Agency, Space Station Utilization Division, Directorate of Manned Space Flight and Microgravity: *The International Space Station, European Users Guide*, Noordwijk, The Netherlands, November, 1998
17. MSS4 Students, "First Draft of the Team Project Literature Review", International Space University, Strasbourg, France, December 1998
18. Eckart, P. *Spaceflight Life Support and Biospherics*, Kluwer Academic Publishers, Dordrecht, 1996

Section Five- Utilization
19. http://mss.isunet/~mss4web/tps/publications/fac_table/index.html

Section Five (Continued)- Technical and Engineering
20. Committee on Use of the International Space Station for Engineering Research and Technology Development, Aeronautics and Space Engineering Board, Commission on Engineering and Technical Systems, National Research Council: *Engineering Research and Technology Development on the Space Stations*, Washington D.C. National Academy Press, 1995
21. European Space Agency, Space Station Utilization Division, Directorate of Manned Spaceflight and Micro-gravity: *The International Space Station: A Guide for European Users* Noordwijk, The Netherlands, BR-137, February 1999
22. European Space Agency, Space Station Utilization Division, Directorate of Manned Spaceflight and Micro-gravity: *The International Space Station: European Users' Guide*, Noordwijk, The Netherlands, November, 1998
23. Messerschmid, E., Reinhold, B., Pohlemann, F.: *Raumstationen: Systeme und Nützung*, Springer Verlag, Berlin, 1997
24. European Space Agency: Exploiting the International Space Station: Research/Services/Education/Commerce, Paris, 1998
25. Kijima, R., Fujimori, Y., Miyama, H.: Perspective of JEM Utilization, *Advances in the Astronautical Sciences* Vol. 91, ppp. 665-674
26. Lecture delivered by Professor Nikolai Tolyarenko.: International Space University, *Lectures on the International Space Station*, Strasbourg, France, 1998
27. National Aeronautics and Space Administration: *The International Space Station: Improving Life on Earth and in Space, The NASA Research Plan, An Overview*, Washington, D.C., 1998
28. Committee on the Space Station, Aeronautics and Space Engineering Board, Commission on Engineering and Technical Systems, National Research Council: *The Capabilities of Space Stations*, Washington, D.C, National Academy Press, 1995
29. Taylor, Lawrence, A.: *He-3 on the Moon: Model Assumptions and Abundances.* 4th Proceedings on Engineering, Construction and Operation in Space, Vol. 1, pp. 678-686, American Society of Civil Engineers, New York, 1994
30. Gazey, S., Kerstein, L. and Apel, U.: *Vision and Long-term Goals for New Business Opportunities in Space*, presented at the 48[Th] Congress of the International Astronautical Federation, Turin, Italy, 1997
31. Toups, Y., and Ximines, S.: *Adaptive Space Station Technologies and Systems for a Lunar Surface Return Mission.* 4th Proceedings on Engineering, Construction and Operation in Space, Vol. 1, pp. 678-686, New York: American Society of Civil Engineers, 1994
32. National Aeronautics and Space Administration: *Potential Pathfinder Areas for Commercial Development of the International Space Station*, http://www.hq.nasa.gov/office/codez/uhran/ppa.html
33. National Aeronautics and Space Administration: Commercial Space Transportation Study, http://stp.msfc.nasa.gov/stpweb/CommSpaceTrans/SpaceCommTransSec1/CommSpacTransSec1.html, December 1998

34. National Aeronautics and Space Administration: *International Space Station: Open for Business*, http://centauri.larc.nasa.gov/issvc97/material.html, December 1998
35. Peacock, A.: *Assembly of a Large X-ray Telescope*, Presented at the Second European Symposium: Utilization of the International Space Station, Noordwijk, Netherlands, 1998

Section Five (Continued)- Life Sciences
36. Presented at the Second European Symposium: Utilization of the International Space Station, Noordwijk, The Netherlands, 1998
37. http://estec.esa.nl/spaceflight/
38. European Space Agency: *Exploiting the International Space Station*: ESA Publications Division, Noordwijk, The Netherlands
39. Ehrenfreund, P., et al.: *Exposure of Organic Matter Onboard ISS*. Presented at the Second European Symposium: Utilization of the International Space Station, edited by Andrew Wilson, ESA Publications Division, Noordwijk, 1998
40. Vergne, J., et al.: *Adenine-Aldehyde Derivatives as Plausible Precursors of Ribonucleosides?"* presented at the Second European Symposium: Utilization of the International Space Station, edited by Andrew Wilson, ESA Publications Division, Noordwijk, 1998
41. Horneck, G.; Exobiology. Life Sciences Research in Space, edited by H. Oser and B. Battrick. ESA Publications Division, Noordwijk , 1989
42. Lecture delivered by Dr. Antonio Guell: International Space University, October 1998
43. Lecture delivered by Dr. Jason Hatten: International Space University, November, 1998
44. Pestov, I.D.: 11th Conference on Space Biology and Aerospace Medicine, Moscow, Russia. June 22-26, 1998 (Information supplied by Dr. Oleg Atkov)
45. http://s2k.arc.nasa.gov/atgb/facts3.html
46. Lecture delivered by Dr. Gerald Soffen: International Space University, *The International Space Station: Improving Life on Earth and in Space, 1998*
47. Roberto, Marco: *Clues From Drosophila To Solve The Paradox: 'Why Microgravity Affects Signal Transduction Pathways In Isolated Culture Cells While Complex Multicellular Organisms Are Able To Develop Unimpaired In The Space Environment?* Presented at the Second European Symposium: Utilization of the International Space Station, Noordwijk, The Netherlands, 1998
48. McLaren, A.: *Developmental Biology. Life Sciences Research in Space*, edited by H. Oser and B. Battrick. ESA Publications Division, Noordwijk, 1989
49. Nechitailo, G. S., and Mashinsky, A. L.: *Space Biology: Studies at Orbital Stations.* Translated by Nikolai Lyubimov, MIR Publishers, Moscow, 1993
50. Moore, D., Bie, P., and Oser, H.: *Biological and Medical Research in Space, An Overview of Life Sciences Research in Microgravity*, p.569, Springer, 1996

Section Five (Continued)- Physical Sciences
51. Houston, A. and Rycroft, M. (editors): *Keys to Space: an Interdisciplinary Approach to Space Studies.* McGraw Hill, New York, 1999

52. Olthof, H.: *Space Science Using the Space Station*, presented at the Second European Symposium: Utilization of the International Space Station, Noordwijk, Netherlands, November 16-18, 1998

53. National Aeronautics and Space Administration: *Space Station Research Payload Utilization Reference Guide*, Houston, ISS Payloads Office, 1996

54. Larter, N. and Gonfalone, A. (editors): *The International Space Station: A Guide for European Users: Early Opportunities Issue*. ESA Publications Division, Noordwijk, Netherlands, 1996

55. European Space Agency: *The International Space Station Microgravity: A Tool for Industrial Research*, 1998

56. Salomon, C. et al.: *Atomic Clock Ensemble in Space on Board the Space Station*, presented at the Second European Symposium: Utilization of the International Space Station, Noordwijk, Netherlands, 1998

57. Huber, F.: *Precision Test of Einstein's Weak Equivalence Principle for Antimatter in Space*, presented at the Second European Symposium: Utilization of the International Space Station, Noordwijk, Netherlands, 1998

58. NASA Official Home Page: http://www.nas.edu/cets/aseb/

59. Microgravity Research Program: *Microgravity News*, Vol. 4, No. 4, Huntsville, Alabama: Marshall Space Flight Center, Winter 1997

60. Personal communication: Alain Berinstain, of the Microgravity Sciences Program, Canadian Space Agency, e-mail of November 30, 1998

61. Canadian Space Agency: http://www.science.sp-agency.ca/K1-MSP(Eng).htm

Section Five (Continued)- Observing Sciences

62. Guyenne, T. D. (editor): *Space Station Utilization Symposium*, ESA Publications Division, Noordwijk, Netherlands, 1996

63. National Aeronautics and Space Administration: *Space Station Research Payload Utilization Reference Guide*, Houston, ISS Payloads Office, 1996

64. Oertel, D. et al.: *FOCUS: Environmental Disaster Recognition System*, presented at the ESA Symposium: Space Station Utilization, Darmstadt, Germany, 1996

65. Mees, J. et al.: *Doppler Wind ALADIN on the International Space Station*, presented at the Second European Symposium. Utilization on the International Space Station, Noordwijk, Netherlands, 1998

66. McCarthy, T.: *Real-Time Mapping using Multispectral Spaceborne Videography Prototype Design for the ISS Encompassing Space and Ground Segment*, presented at the Second European Symposium: Utilization of the International Space Station, Noordwijk, Netherlands, November 16-18, 1998

67. Kaye, J. A.: *NASA's Plans for Earth Science Research from the International Space Station*, presented at the Second European Symposium. Utilization on the International Space Station, Noordwijk, Netherlands, November 16-18, 1998

68. NASDA Official Home Page: http://jem.tksc.nasda.go.jp/JEM/jemmefc/english/smiles.html

69. Kondo, Y.: (editor): *Observatories in Earth Orbit and Beyond*, Kluwer Academic Publishers, Dordrecht, 1990

70. Wertz, J. R. and Larson, W. J. (editors): *Reducing Space Mission Cost*, Microcosm Press and Kluwer Academic Publishers, Torrance, USA, and Dordrecht, The Netherlands, 1996

Section Six- How to Utilize the ISS

71. Leuttgens, R. and Volpp, J.: Operations Planning for the International Space Station, *ESA Bulletin*, pp.57-63, May 1998
72. NASA JSC MOD Space Flight Training Division: *International Space Station Familiarization*, TD9702, December 1997
73. cfr. NASA-ESA MoU art. 8.1.c
74. cfr. NASA-ESA MoU art. 8.3.g.3
75. European Space Agency: *The International Space Station — European Users Guide*, pp. 49-54, November 1998
76. European Space Agency: International Life Science Research Announcement (LSRA), SP-1210, December 1996
77. National Aeronautics and Space Administration: *Potential Pathfinder Areas for Commercial Development of the International Space Station (Draft)*, Washington, D.C., October 1998
78. National Aeronautics and Space Administration: *Commercial Development Plan for the International Space Station*, Washington, D.C., November 1998
79. National Aeronautics and Space Administration: ISS — *The NASA Research Plan, an Overview*
80. National Aeronautics and Space Administration: *A Non-Government Organization (NGO) for Space Utilization Management (Draft)*, Washington, D.C., October 1998
81. NASA-ESA MoU, Early Utilization Opportunities of the International Space Station
82. Wetter, B.L.: *Canadian Programme and Research plans for International Space Station Utilization*, Proceedings of the 2nd European Symposium on the Utilization of the International Space Station, ESTEC, Noordwijk, The Netherlands, November 16-18, 1998, pp. 41-45 (ESA SP-433, February 1999)

The information compiled within this User's Overview has been researched, analyzed and produced by the following individuals:

The Class- MSS 4 (1998 – 1999)

Rob Alexander, USA
Youssef Attia, Libya
Eugeniu Caisin, Moldova
Lokman Dagli, Turkey
Mosbah Elkhrad, Libya
Simone Garneau, Canada
Ivan Gracnar, Slovenia
Özgür Gürtuna, Turkey
Claire Jolly, France
C.P. Karunaharan, England
Udo Kugel, Germany
Vaios Lappas, Greece / Canada
André Larisma, South-Africa / Portugal
Anders Lindsköld, Sweden
Suparna Madhu, India
Tarek Melad, Libya
Martha Milkeraitis, Canada
Khalid Musa, Libya
Tomofumi Ono, Japan
Andrew Ray, Canada
Karin Remeikis, Germany
Claude Rousseau, Canada
Yakov Sadchikov, Kazakhstan
Marek Sadowski, Poland
David Sylvester, USA
Yoshimasa Tajima, Japan
Munir Tarar, Pakistan
Stella Tkatchova, Bulgaria

Editorial Task Implementation Group

Edoardo Benzi, Italy
Bill Boardman, USA
Tare Brisibe, Nigeria
Ruofei Gao, China
Laurance Higgs, England
Caroline Maredza, Zimbabwe
Jake Maule, England
Piero Messina, Italy
Raman Mittal, India
Mehrdad Rezazad, Iran

Appendix 1   Final Assembly ISS Exploded Diagram

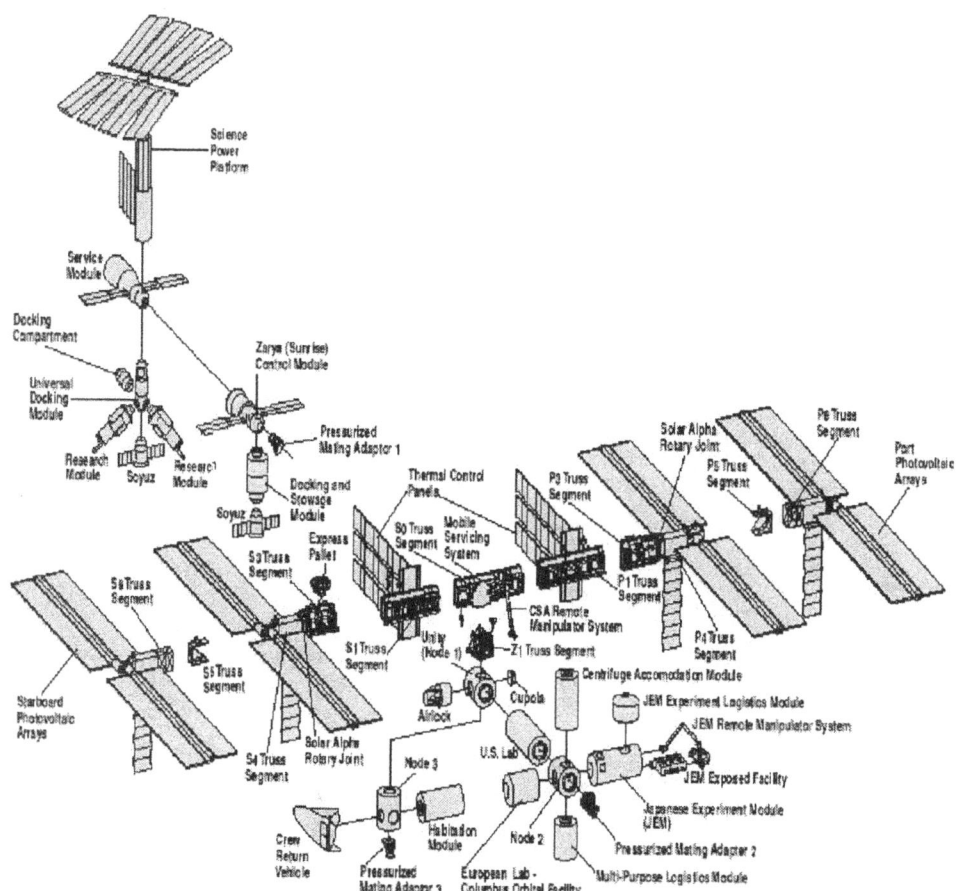

Appendix 2   Precis of Information on the ISS

Comprehensive and useful data on the ISS facilities are provided on the web site:

http://mss.isunet.edu/~mss4web/tps/publications/fac_table/index.html

These data consider the

- Function
- Location
- Maximum workspace (volume or mass) available for experiments
- Maximum and minimum microgravity level anticipated
- Diagnostic or analysis tools available
- Accuracy of positioning or orientation
- Interfaces and operation modes
- Crew time or experiment time
- Available resources (power, data, communications, working fluids)
- Assembly date
- Announcement of Opportunity date
- Priority level
- Jurisdiction for patent filing and Intellectual Property Rights
- Responsibility for access
- Comments on compatibility with other payloads
- Contact addresses

for

Standard Payload Accommodation Facilities

- International Standard Payload Rack (ISPR)
- Express Pallets (EP)
- Express Rack (ER)
- European Drawer Rack (EDR)

Fluid Combustion Facility (FCF)

- Fluid Physics Experiment Facility (FPEF)
- Combustion Rack
- Fluids Rack
- Fluid Science Laboratory (FSL)

Material Science Facilities

- Gradient Heating Furnace (GHF)
- Advanced Furnace for microgravity Experiment with X-ray radiography (AFEX)
- Electrostatic Levitation Furnace (ELF)
- Low Temperature Microgravity Physics Facility
- Materials Sciences Research Facility (MSRF)
- Materials Sciences Laboratory (MSL)
- Advanced TEMPUS
- Solution/Protein Crystal Growth Facility (SPCF)
- Protein Crystal Diagnostic Facility (PCDF)
- Advanced Protein Crystallization Facility (APCF)
- Biotechnology Facility (BTF)

Life Sciences Facilities

- Human Research Facility (HRF 1 & 2)
- Biolab
- European Physiology Module (EPM)
- Centrifuge Facility
- Life Sciences Glovebox (LSG)
- Gravitational Biological Facility (GBF)
- Modular Cultivation Systems (MCS)
- Minus 80 degree Laboratory Freezer for ISS (MELFI)
- Cryogenic Storage Freezer (CSF)
- Quick/Snap Freezer (QSF)

Exposed Facilities

- Express Pallets (EP)
- Technology Exposure Facility
- JEM Exposure Facility (JEM-EF)
- Space Exposure Biology Assembly (SEBA)

Observation Facility

- Cupolas
- Windows Observation Research Facility

Appendix 3   Acronyms List

AAEF	Aquatic Animal Experimental Facility (in JEM module)
ABDM	Advanced Bone DensitoMeter
ACES	Atomic Clock Ensemble in Space
ACS	Atmospheric Control and Supply
ADAS	Ambulatory Data Acquisition System
AFEX	Advanced Furnace for microgravity Experiment with X-ray radiography
ALADIN	Atmospheric Laser Doppler Instrument
AMS	Alpha Magnetic Spectrometer
AO	Announcement of Opportunity
APCF	Advanced Protein Crystallization Facility
APM	Attached Pressurized Module
AR	Atmosphere Revitalization
ARMS	Advanced Respiratory Monitoring System
ASCR	Assured Safe Crew Return
ASI	Agenzia Spaziale Italiana (Italian Space Agency)
ATV	Automated Transfer Vehicle
BDM	Bone DensitoMeter
BIOLAB	Biological Laboratory
BIOPAN	(Exposure platform, shaped as a circular container)
BIVOG	Binocular VideoOculGraph
BMAS	Biomedical Analysis System
BSMD	Bone Stiffness Measurement Device
CADMOS	Centre d'Aide au Developpement de la Micropesanteur et aux Operations Spatiales
CAM	Centrifuge Accommodation Module
CB	Clean Bench (in JEM module)
CBEF	Cell Biology Experimental Facility (in JEM module)
CBPD	Continuous Blood Pressure Device
CCD	Charge Coupled Device
C&DH	Command and Data Handling
CdTe	Cadmium Telluride
CF	Centrifuge Facility
CFZF	Commercial Float Zone Furnace
CHeCS	Crew Health Care System
CNES	Centre Nationale d'Etudes Spatiales (France)
COF	Columbus Orbital Facility
COF-CC	COF Operation Center
COUP	Consolidated and Utilization Plan
CRV	Crew Return Vehicle
CSA	Canadian Space Agency
CSF	Cryogenic Storage Freezer

CSC	Commercial Space Centers
C&TS	Communication and Tracking System
CUP	Common Utilization Plan
DARA	Deutsches Agentur fur Raumfahrtangelegenheiten (German Space Agency)
DRTS	Data Relay Test Satellite
EC	European Community
ECG	Ambulatory Electrocardiogram
ECLSS	Environmental Control and Life Support System
EDR	European Drawer Rack
EEG	Electroencephalogram
EFA	Engineering Feasibility Assessments
ELITE	Motion analysis system in European Physiology Module
EMIR-2	European Microgravity Research
EMRS	Eye Movement Recording System
EPF	External Payload Facility
EPM	European Physiology Module
EPS	Electrical Power System
EPS	European Partner State
ERA	Exobiology and Radiation Assembly
ESA	European Space Agency
ESS	Eye Stimulation System
EUB	European Utilization Board
EVA	Extra Vehicular Activity
FDS	Fire Detection System
FGB	Functional cargo Block (= "Zarya")
FOCUS	Fire Detection Infrared Sensor System
FRC	Facility Responsible Centers
FSL	Fluid Science Lab
GASMAP	Gas Analyzer for Metabolic Analysis of Physiology
GBF	Gravitational Biology Facility
GHF	Gradient Heating Furnace
GN&C	Guidance, Navigation and Control
GOJ	Government of Japan
GPS	Global Positioning System
GTS	Global Transmission Service
HAB	U.S Habitation module
HGD	Hand Grip Dynamometer
HIPE	High-resolution Photogrammetric Experiment
HRD	High Rate Data
HRF	Human Research Facility
HRMRB	Human Research Multilateral Review Board
HTV	H-IIA Transfer Vehicle
IAS	Internal Audio Subsystem
IDRD	Increment Definition and Requirements Document

IEPT	International Executive Planning Team
IGA	Inter-Governmental Agreement
IMBP	Institute of Medical and Biomedical Problems
IMSPG	International Microgravity-science Strategic Planning Group
IP	Intellectual Property
IPR	Intellectual Property Rights
ISPR	International Standard Payload Racks
ISS	International Space Station
ISSLSWG	International Space Station Life Science Working Group
ISU	International Space University
ITA	Integrated Truss Assembly
IVA	Intra Vehicular Activity
JEM	Japanese Experiment Module
JEM-EF	Japanese Experiment Module Exposed Facility
LAN	Local Area Networks
LBNP	Lower Body Negative Pressure
LEO	Low Earth Orbit
LSG	Life Sciences Glovebox
LSRA	Life Science Research
MAP	Microgravity Applications Promotion
MARS	Microgravity Advanced Research and Support Center
MBS	Mobile remote service Base System
MCB	Multilateral Coordination Board
MCC-H	Mission Control Center - Houston
MCC-M	Mission Control Center - Moscow
MELFI	Minus Eighty Degree Laboratory Freezer for the ISS (cryosystem)
MFC	Microgravity Facility for Columbus
MIM	Microgravity vibration Isolation Module
MIP IPT	Multi-Increment Planning Integrated Product Team
MITI	Ministry of International Trade and Industry (Japan)
MOU	Memorandum Of Understanding
MPLM	Multi Purpose Logistics Module
MSFC	Marshall Space Flight Center
MSL	Material Science Lab
MSRRs	Materials Science Research Racks
MSS	Master of Space Studies
MUSC	Microgravity User Support Center
NASA	National Aeronautics and Space Administration (USA)
NASDA	National Space Development Agency of Japan
NGO	Non-Governmental Organization
OES	Office of Earth Science
OFA	Operational Feasibility Assessment
OLMSA	Office of Life & Microgravity Science and Applications
OOS	On-orbit Operational Summary

PCDF	Protein Crystal Diagnostics Facility
P/L	Payload
POIC	Payload Operations Integration Center
PP	Planning Period
PPS	Planning Period Start
PSOs	Protected Space Operations
PUP	Partner Utilization Plan
QSF	Quick/Snap Freezer
R&D	Research and Development
RKA	Russian Space Agency
ROS	Russian Orbital Segment
RSA	Russian Space Agency
SAGE	Stratospheric Aerosol and Gas Experiment
SEBA	Space Exposure Biology Assembly
SECAM	Sequential Couleur a Memoire
SESAM	Surface Effect Sample Monitor
SIRs	Stage Integration Reviews
SLAMMD	Space Linear Acceleration Mass Measurement Device
SMILES	Superconducting subMIllimeter Limb Emission Sounders
SOAR	Station Off-Axis Rotator
SOP	System Operations Panel
SPOrt	Sky Polarization Observatory
SSCC	Space Station Control Center
SSRMS	Space Station Remote Manipulator System
STP	Short-Term Plans
STS	Space Transportation System (e.g. Space Shuttle)
TBD	To Be Decided
TCS	Thermal Control System
TDRS	Tracking and Data Relay Satellite
TDRSS	Tracking and Data Relay Satellite System
THC	Temperature and Humidity Control
UFs	Utilization Flights
UHF	Ultra High Frequency
ULC	Unpressurized Logistics Carrier
UOP	User Operations Panel
US	United States
USA	United States of America
USOS	United States Orbital Segment
USSR	Union of the Soviet Socialistic Republics
UTEF	Ultraviolet Telescope Facility
UV	Ultra Violet
VDS	Video Distribution Subsystem
VEG	Virtual Environment Generator
WEP	Weak Equivalence Principle

WRM	Water Recovery and Management
XEUS	X-ray Evolving Universe Spectroscopy
ZOE	Zone of Exclusion

SPACE STUDIES

KLUWER ACADEMIC PUBLISHERS – DORDRECHT / BOSTON / LONDON

The manufacturer's authorised representative in the EU is Springer Nature Customer Service Centre GmbH, Europaplatz 3, 69115 Heidelberg, Germany. If you have any concerns regarding our products, please contact ProductSafety@springernature.com

Printed and bound by CPI Group (UK) Ltd, Croydon, CR0 4YY

23/04/2026

02095628-0003